europe's **leading** guide to the ferry industry

contents

Above: **Visby** *(Miles Cowsill)*
Frontispiece: **Dronning Margrethe II** *(John May)*

europe's **leading** guide to the ferry industry

introduction

This is the eighteenth edition of this book, which first appeared in 1983 as the 24 page 'home published' *'Car Ferries from Great Britain and Ireland'*. The book aims to list every passenger/vehicle ferry in Great Britain and Ireland, ro-ro freight vessels which operate regular services between Great Britain and Ireland and to nearby continental destinations and major passenger/vehicle ferries in other parts of Northern Europe. The coverage of Northern Europe is not fully comprehensive (to make it so would probably triple the size of the book) and does not include freight-only operations and vessels- although for the first time some freight-only vessels have been included where the operators also run passenger services. Also, ro-ro vessels engaged in 'deep sea' and Mediterranean trade and those operated solely for the carriage of trade cars are not included.

Each operator is listed alphabetically within sections – international and Northern Ireland routes, domestic services, freight-only operations, chain, cable and float ferries, passenger only ferries, other North European passenger operators and vehicle/passenger vessels owned by companies not currently engaged in operating services. After details relating to each company's management, address, telephone numbers, email, web site and services, there is a fleet list with technical data and then a potted history of each vessel with previous names and dates. For the first time I have listed the shipyard in which the vessel was built.

2004 saw the delivery of two exciting new passenger vessels – Color Line's *Color Fantasy* and Brittany Ferries' *Pont-Aven* – confounding those who assert that the age of the cruise ferry is gone. A new Tallink vessel and an even larger sister to the Color Fantasy are promised for the future. On the freight side, there have also been impressive new deliveries such as DFDS Tor Line's 'Flower Class' and new ships are on order for Stena Line, Cobelfret Ferries and Norfolkline. However, these are all large ships for major companies and cannot hide the chronic shortage of medium sized vessels to replace the many 1970s built vessels which are rapidly reaching the end of their lives. Even those companies who operate ships for charter only seem to be ordering large ships for companies who can give them a guaranteed lengthy initial charter such as Transfennica. Unless someone starts building ships more appropriate for smaller and start-up companies and more marginal routes, there will be major problems ahead.

Whitstable, Kent **Nick Widdows**

May 2005

europe's **leading** *guide to the ferry industry*

foreword

Søren Jespersen

It is with great pleasure that I write a foreword to this practical reference guide, which helps keep all of us working within the industry abreast of current changes. Indeed, change is a constant uniting our world and businesses. Presently the dramatically lower cost of airfares continues to add competitive pressure on the business of passenger shipping, and within just a brief period of time airline companies have in effect changed the way Europeans travel. Lower ticket prices combined with greater affluence has enabled more individuals and families to leave everyday life and stress behind and take short holiday breaks abroad. So while it is true that more people head for the airport, a greater number of people are also in the market to get pampered, celebrate and spend a good time together.

So where does passenger shipping fit into this new context? In all honesty there are more luxurious hotels, wider choices of restaurants, more select shopping opportunities, cheaper and certainly faster means of transport to be found. But rather than entering into a head-on price war as practised by airline companies, a change of focus is needed to counter competition; from transport to experience, from purely price-sensitive destination sales to emphasis on the high quality cruise experience. Because being at sea is something else.

During the second half of 2004 DFDS Seaways began work on our new Mini Cruise concepts. It is an exciting time for us. Detailed customer segment analyses have been conducted in all our markets providing us with vital information about our existing and potential guests. Based on this a new marketing platform and upgraded Mini Cruise concepts have been developed and finalised. In early 2005 they were introduced in the Danish, Norwegian and Swedish market, helping us cut a sharper profile in the competitive sector for brief getaways and allowing us to attract a wider audience of guests looking to enjoy themselves.

The maritime experience forms the unique core of the concept, and while this represents a breath of fresh air on board, it also brings us closer to the heart of our business. Guests encounter our maritime story in both the hardware and the on-board activities and facilities, designed to not only meet, but also exceed the wishes of all our different guests, private as well as business, young as well as old. Yet above all, delivering a high quality Mini Cruise experience is a people to people business. DFDS Seaways already uphold a reputation for high service, and our staff have been involved right from the start. Almost 700 members of staff, from terminal workers to captains, have been further trained in delivering consistent, outstanding service to our guests, and many new service and experience improvements are the result of their suggestions. Change breeds change.

I am sure that you will find this book a useful tool throughout 2005, because fortunately some things don't change, and on behalf of DFDS Seaways I hope to welcome you aboard to an upgraded cruise experience very soon.

Søren Jespersen,
DFDS Seaways

Pearl of Scandinavia *(Miles Cowsill)*

Crown of Scandinavia *(Miles Cowsill)*

dfds
a brief history

DFDS, Det forende Dampskibs-Selskab – in English The United Steamship Company – even the name conjures up visions of a glorious past for a ship owner celebrating its 139th birthday this year. From the early vision of C F Tietgen, the architect of what was to become one of the largest shipping companies in the world, has evolved the DFDS that we see today with its fleet of passenger ships and impressive modern freight ferries; a company that is still Master of the Northern Seas, albeit in the early part of the 21st century in freight operations rather than the passenger shipping for which the phrase was originally coined. Danish financier, Carl Frederik Tietgen, born in Odense in 1829, had worked hard to become a leading merchant in Copenhagen by the mid 1850's. In 1857 he became the first director of a new bank in the Danish capital, set up by some of that city's leading financiers.

Tietgen's attention was first drawn to the shipping industry in 1865, when a failed deal left four ships uncompleted in a British shipyard. In order to raise the capital to complete these vessels he planned the merger of three of the largest Danish ship owners into a new enterprise. The oldest of these companies was H P Prior, founded in 1850, bringing nine ships. The others were A/S Det Almindelige Danske Dampskibs-Selskab and Koch & Henderson, an Anglo-Danish operation. Officially formed on 11th December 1866, DFDS began trading on 1st January 1867 with a fleet of 19 ships serving Danish domestic, North Sea and Baltic routes. The first acquisition by the new company was a second-hand iron-hulled steamer of 150 gross tons, the *St Knud*. The first new ship was the cargo paddle steamer *Dania*, ordered to replace a ship of the same name owned by H P Prior, but lost just before the merger. By 1872, by means of acquisition, DFDS controlled most of Denmark's domestic steamship trade. Further companies were acquired over the next few years, including a company operating various routes between Denmark and Sweden across the Sound.

Expansion was not just about buying up other ship owners, and the company actively looked for new routes. In 1874 the service from Stettin to Copenhagen and on to what we now know as Oslo was commenced, and in the following year Esbjerg to Thameshaven (later the English terminal was changed to Harwich) sailings began. By 1887 DFDS had expanded its route network to include services to the Iberian Peninsula, the Mediterranean and the Black Sea. The latter route was, perhaps, the company's most impressive at the time, beginning in St. Petersburg and ending at Odessa. In these early days the company carried a good mix of passengers, livestock, general cargo and dairy products on a variety of ships, all purpose built for their particular trade. Until 1878 new cargo ships had been screw driven and passenger ships had been predominantly paddle steamers. The last paddle steamers to be ordered were the *H P Prior*, delivered in that year for the Aalborg to Copenhagen service, and the *Koldinghuus* of 1883 for the Esbjerg to Harwich route.

A milestone in the history of the company, by now one of the ten largest ship owners in the world, occurred in 1894 when it branched out across the Atlantic Ocean to link Copenhagen with New Orleans. With an obvious liking for trans-Atlantic business, in 1898 DFDS bought the established operator A/S Dampskibsselskabet Thingvalla, together with its four ships employed on the route linking Scandinavian ports with New York. That company was restyled as Scandinavia-America Line and three 10,000 gross ton ships were ordered from the Glasgow shipyard of Alexander Stephen & Son. These were by far the largest ships to be owned by the company to date, and the first of these, the *Oscar II*, entered service in 1902 with a capacity for 1,100 passengers. Around this time, the Black Sea route was discontinued and the ships transferred to the New Orleans service. The first *C F Tietgen* was built as the *Rotterdam* for Holland America Line in 1897. She was acquired by DFDS in 1906 and immediately entered into the New York trade. The *Frederik VIII*, then the largest Scandinavian owned ship at 11,850 gross tons,

Winston Churchill *(Bruce Peter)*

Dana Gloria *(FotoFlite)*

replaced her in 1913. The New York service eventually proved to be a profitable route, surviving until 1935, by which time the USA had seriously cut back on immigration quotas.

A new century and new opportunities; in 1906 the Black Sea service from the Baltic to Romanian ports was reintroduced and during the following year a freight route between Northern Europe and Argentina (later including calls in Brazil) was established to send coal south and to return with coffee and grain. An unusual purchase in 1915 was the four-masted barque *Viking*, which the company used as a sail training ship until sold on in 1929. This ship, probably the oldest DFDS-owned vessel still afloat can be seen today in Gothenburg.

In 1916 DFDS acquired control of the Elsinore Shipyard, which was to build more than 80 ships for the company. The company had some experience of the shipbuilding industry due to its participation in the financial reconstruction of the Frederikshavn Yard three years earlier. Despite Danish neutrality during the First World War, the company lost 26 ships as a direct result of acts of war, and a further five were wrecked. With the end of the war a programme to replace the losses was put in place and the fleet was also expanded to take full advantage of the increase in trade. However, the upturn was short-lived, and by 1921 the freight market was in a state of near-collapse, resulting in up to 30 DFDS ships being laid up for varying periods. In an attempt to counter this decline, the company opened up new routes, including Esbjerg to Antwerp and Dunkerque, and Copenhagen to Leningrad. In 1933 the fleet peaked at 123 ships of various types, including transatlantic liners, passenger ferries, freighters of many types and eight tugs. The first *England*, a beautifully profiled motor ship of 2,700 gross tons, had been delivered the year before for the Esbjerg to Harwich service to supplement the older but similarly sized *Esbjerg* and *Parkeston*. Although these ships represented the pinnacle of pre-Second World War North Sea vessels, the *Kronprins Olav*, dating from 1937 and built for service between Copenhagen and Oslo, was undoubtedly the finest ferry that the company had owned to date. She presented an even more attractive profile after a major re-build in 1950. It is not unreasonable to conclude that a deliberate policy of building passenger ferries of outstanding appearance was instigated in the late 1920s, and continued for almost 50 years, concluding with the arrival of the *Dana Regina* in 1974.

In 1937 the route from Copenhagen to the Levant was reintroduced and two years later passenger sailings, albeit in a very limited form, recommenced between Copenhagen and New York using four freighters with capacity for twelve passengers. That service did not last for very long and was discontinued when Germany occupied Denmark in April 1940.

The Second World War cost the company dearly, with 24 ships lost due to enemy action and a further seven lost by other means. The British requisitioned 21 ships, and another five were eventually to find war service under United States control. Occupying forces took control of many other vessels, leaving the company with just a few ships to serve domestic and Baltic routes.

Post-war fleet re-building commenced in 1945 and gradually the routes that had been disrupted by the war were reinstated. The first new ship to enter service was the beautiful *Kronprins Frederik*, which although almost completed in 1941 was laid up for the duration of the war with vital engine parts missing so that she could not be used by the occupying forces. In the five years to 1951 the company took delivery of 27 cargo ships and the passenger ferries *Kronprinsesse Ingrid* (Esbjerg-Harwich) and the *Jens Bang* and *H P Prior* (both Copenhagen-Aalborg). The company then operated a total of 89 ships, still including eight tugs. The Esbjerg to Newcastle route was established in 1953.

In what was widely regarded as an irregular and possibly illegal series of share purchases, Danish shipowner J Lauritzen seized control of DFDS in 1964, although it continued to run the company as a separate entity. By 1966, the company's centenary year, the fleet had shrunk to 70 vessels. The main passenger routes were those from Copenhagen to Oslo, Århus and Aalborg, from Esbjerg to Harwich and to Newcastle, and Frederikshavn to Oslo. The company's freight ships were still engaged on trades throughout Western Europe, the Mediterranean and the Americas.

The last major passenger ship construction programme began in 1967, comprising of the *Winston Churchill* (Esbjerg-Harwich), the *Kong Olav V* and *Prinsesse Margrethe* (both Copenhagen-Oslo), and the *Aalborghus* and *Trekroner* (both Copenhagen-Aalborg). Additionally, the freight ro-ro's *Somerset*, *Stafford* and *Surrey* entered service on the North Sea. Domestic passenger services ceased in September 1970 (domestic freight services closed a year later) and in conjunction with the start of ferry services in the Mediterranean, DFDS Seaways became the marketing name for all of the passenger

Dana Anglia *(DFDS)*

operations. The *Aalborghus* was re-built at Marseilles for her new service between Genoa and Palma/Malaga during the early part of 1971, eventually entering seasonal service in June of that year as the *Dana Sirena*. She subsequently also served the ports of Patras, Bari and Tunis until withdrawn and chartered out at the end of the 1975 season. The *Trekroner*, as the *Dana Corona* (and later renamed *Dana Sirena*) served similar Mediterranean routes until the company withdrew from the region in 1982.

The late 1970s saw the start of DFDS's shift of emphasis from passenger/car ferry to freight ro-ro, and that decade also signalled the end of the company's conventional cargo services. The acquisition of a number of British and Danish road hauliers into what was to become DFDS Transport, a name embracing all of the freight operations eventually, strengthened the move towards freight ro-ro operations. The road haulage business was disposed of around the turn of the century as the company concentrated its attention once again on ships. The *Dana Regina* replaced the *England* on the Esbjerg to Harwich service in 1974, being joined by the *Dana Anglia* four years later. The *Dana Anglia* was the last passenger/car ferry to be built for any of the passenger services of the company, although there have been a number of ships acquired subsequently.

The early 1980s were a time of expansion, following the retrenchment of the previous decade. In 1981 Prinzenlinien together with its route between Harwich and Hamburg was absorbed, and during the following year the Swedish ro-ro operator Tor Line joined the DFDS group, bringing the magnificent greyhounds *Tor Britannia* and *Tor Scandinavia* to the 'Seaways' fleet and a network of North Sea routes to the freight operation. The next passenger ship to join the fleet was the former Silja Line vessel, *Wellamo*, renamed *Dana Gloria* for Newcastle services initially, and later used on the Copenhagen to Oslo route. Lengthened in 1989, she was renamed *King of Scandinavia* and was later transferred to Color Line to become the *Color Viking*. In conjunction with the lengthening of the 'King', another Silja Line ship, the *Finlandia*, was purchased in 1990 for use on the Oslo service as the *Queen of Scandinavia*.

The ill-fated venture into the North American cruise ferry operation was short-lived, and caused very serious financial problems for the company. A two ship service between Florida and the Bahamas was introduced in 1982, but the jewel in crown, the 40,000 gross ton *Scandinavia*, delivered new later that year, opened a new service between New York and Freeport in the Bahamas. That service survived for

Dana Sirena *(Andrew Cooke)*

just one year, but it took DFDS rather longer to extricate itself fully from the disastrous North American cruise venture. As a temporary move, the *Scandinavia* was relocated to the Oslo service. A little more that a year later she was sold and renamed *Stardancer*. She currently operates for First Choice Holidays as the *Island Escape*.

The Hamburg service was upgraded in 1987 with the arrival of the *Hamburg*, built for Jahre Line as the *Kronprins Harald* in 1976. Subsequently renamed *Admiral of Scandinavia*, she was a popular ship on the German route. She was sold in 2002 and replaced by a vessel with a rather smaller passenger capacity as the company strove to fight back against declining passenger traffic occasioned by the rapid growth of low cost airlines. From 1985, the passenger ferries began to adopt the new 'blue stripe' livery, with the operating name being changed to Scandinavian Seaways three years later. The ships were renamed with 'of Scandinavia' names, the *Tor Britannia* becoming 'Prince' and the *Tor Scandinavia* 'Princess'. With declining passenger numbers on most services in the North Sea, there has been little incentive to modernise what is now an ageing fleet, and routes such as those from Newcastle to Esbjerg and Hamburg were abandoned. The bright spots on an otherwise gloomy picture have been the continued renewal of ships on the Capital Cities route linking Copenhagen with Oslo, and the introduction of the extremely successful service from Newcastle to Amsterdam (IJmuiden), and a gradual, if less spectacular upgrading of tonnage to provide a daily service.

The early years of the current decade have seen a massive investment in new and fast freight ferries as the so-called 'Flower Class' nears the end of its delivery programme. The company has also disposed of a number of older units, some of which are still operated under a charter-back arrangement. DFDS now operates an impressive network of Baltic and North Sea freight routes under the style DFDS Tor Line, and although on the Northern Seas freight is now king, it is be hoped that passengers continue to use DFDS's services in sufficient numbers to enable renewal of the older units and a continuance of the long tradition of high quality service for which the company is rightly known.

William Mayes

crown **and** pearl of scandinavia

dfds's famous oslo boats

The overnight voyage from Copenhagen to Oslo is one of the most genteel and delightful journeys by ferry one can make in Northern Europe. Leaving central Copenhagen, one first enjoys the picturesque coastal scenery of The Sound from a high vantage point on the ship's sun deck. After a couple of hours, a brief call is made at the busy port of Helsingborg to pick up Swedish passengers, and the ship has carefully to negotiate the busy traffic of car and train ferries shuttling backwards and forwards on the short 20-minute crossing between Helsingør and Helsingborg. As the sun sets in the West, the medieval ramparts and later Renaissance spires of Kronborg Castle are outlined on the horizon. Traditionally, DFDS's ships on the Copenhagen-Oslo route have been amongst the finest and most innovative of their type. In the mid-1930s, the company introduced its splendidly streamlined *Kronprins Olav*, which brought advanced Modernist design to Scandinavian shipping for the first time. Later, in the post-war era, the *Prinsesse Margrethe* and *Kong Olav V* of 1957 and 1961 added further refinements and were highly acclaimed for their elegance and luxury. Moving into the car ferry age, two new ships of the same names, introduced in 1968, were equally luxurious – and even more spacious. These lasted until the early 1980s when the North Sea vessels *Dana Regina* and *Dana Gloria* were introduced. Briefly, in 1984-5, the giant cruise ferry *Scandinavia* was tried with great success. This voluptuous vessel had tiered show lounges, swimming pools, Jacuzzis, nightclubs and every modern refinement expected of the latest American-based cruise liners (it had been built to sail between New York and the Bahamas) – and passenger numbers soared. But, ultimately, the ship proved too costly and was replaced by more moderate tonnage.

In the early 1990s, DFDS once again felt that it could afford cruise ferries for the route and so it introduced the former Silja Line ferry *Finlandia* as the *Queen of Scandinavia*. Later, in 1994, the brand new *Crown of Scandinavia* was acquired fresh from the builder and, later still, the *Pearl of Scandinavia* was purchased. These ships are full of wonderful facilities and so, once passengers have retired indoors for the evening, they are able to enjoy wide ranges of dining and entertainment options. The next morning, they awaken to the delight of a slow cruise up the Oslofjord, culminating in a grand panoramic vista across the expansive harbour of the handsome Norwegian capital, punctuated by Akershus Castle and by the impressive City Hall.

Not only has the Copenhagen-Oslo route used some of the best ships, but also it has carried many distinguished guests over the years – including senior members of the Danish and Norwegian royal families. When the 'Oslo boats' used to dock at Larsens Plads in Copenhagen, adjacent to Amalienborg Palace, the Danish King, Frederik IX, who was a shipping and railway enthusiast, was known to telephone the DFDS headquarters at Kvæsthusbroen to enquire why the morning arrival from Oslo was even slightly delayed! Although the present ships do not have 'royal' names in the true sense, a great deal of heritage and prestige remains with the route today.

THE *CROWN OF SCANDINAVIA*

The *Crown of Scandinavia* is the last of four similar sister ships built by the Brodogradevna Industrija shipyard at Split in Croatia in the 1988-1993 period. Viking Line's *Amorella* and *Isabella* were the first of this class. The latter pair were improved versions and were intended for a service from Sweden to Germany, marketed as Euroway. Thus, they were fitted with more overnight capacity in cabins than their Viking Line sisters, some of which are located atop the superstructure and even featured verandas. The early-90s recession was particularly cruel to Sweden (a country which then had a heavy debt burden), and the service was unsuccessful. Only the first sister – the *Frans Suell* – was used but the second – to have been named the *Thomas Mann* – was sold instead to DFDS and completed by Lloyd Werft in Germany, before entering the Copenhagen-Oslo service in 1994. The

Crown of Scandinavia *(Bruce Peter)*

Columbus Bar - Crown of Scandinavia *(DFDS)*

Frans Suell later became the *Silja Scandinavia* and is now Viking Line's *Gabriella.*

Upon acquiring the *Crown of Scandinavia,* DFDS carried out many alterations and improvements before she entered service. A new marble-floored reception hall two decks below the original was created to fit its gangway arrangements in Copenhagen and a duck-tail sponson was added to the stern. All of the interiors are the work of the Swedish architect Robert Tillberg and are very similar to his designs for the ships's near sisters in the Viking Line fleet.

I first visited her in Copenhagen when newly delivered and have since made several enjoyable crossings on her. During her layover, she was being given the most thorough clean by a large and enthusiastically attentive Danish crew. Florists were hard at work placing new arrangements in the foyers and public rooms.

A lift whisked me up to the ship's topmost deck and her penthouse suites, where the tour began. These are as commodious as many staterooms on recent cruise liners and the centrepiece of this deck is an airy garden lounge with a large, domed skylight, where one can sit with a good book and drink a cup of coffee away from the throng of shipboard life (equally, one could relax on one's private balcony, but not in the freezing winter weather).

Moving down to the main public rooms a few decks below, an attractive arcade on the starboard side links these and makes orientation very simple. Forward is the Seven Sea Buffet Restaurant, which has a long, centrally-located servery and a series of smaller seating spaces on either beam. There are floor-to-ceiling windows in the sharply-raked forward superstructure. Having a table here on a summer evening as one sails down the Oslofjord is a memorable experience. Moving aft, one comes to two further restaurants – a steakhouse, inboard of the arcade and the Blue Ribband A La Carte. Each is typically 'Tillberg' with glass chandeliers, ornately-patterned moiré silk wallpaper behind glass panels and Gustavian furniture. Part of the Blue Ribband Restaurant has long tables for party and conference groups to dine and make merry together – an important aspect of the Copenhagen-Oslo route. Adjacent to this is a small wine and cigar bar.

Returning to the arcade and heading aft, one next comes to an 'English Pub' and, filling the entire aft section are the main entertainment spaces. The Columbus Club is a big lounge with three bars and a circular dance floor with a bandstand behind on the port side. One deck above is the two-level discotheque which is well located and has a marvellous double-height central section with complex sound and lighting effects.

One deck below the main saloon deck is the original circular entrance hall with reception desk (used in Oslo), the cafeteria and the children's playroom. Cabins are located fore and aft. Continuing further down, one comes to further decks of cabins and the swimming pool, sauna and fitness centre, located in the ship's bow. Overall, the Crown of Scandinavia is a superbly impressive ship. Even after ten years' of usage, she still looks split new and everywhere one looked the maintenance was very thorough. DFDS, her captain and crew are obviously very proud of her.

THE *PEARL OF SCANDINAVIA*

Although the most recent ship to be introduced on the Copenhagen-Oslo route, the *Pearl of Scandinavia* has had a much more complex history. While the *Amorella* and the *Isabella* were taking shape in Yugoslavia, the Swedish partner in the Viking Line consortium, Rederi AB Slite, had ordered two new ferries of its own from the Wärtsilä shipyard in Helsinki. The first of these 40,000t giants, for delivery in 1989 and to be named the *Athena,* was actually intended to be used as a cruise ship, replacing the steamer *Apollo III* (ex *Svea Jarl*) on the Stockholm-Mariehamn route. Unfortunately, during the construction of the *Kalypso,* the Wärtsilä shipyard became financially stretched and, indeed, it nearly went bankrupt. In order to secure their investments, the least worst option for the shipping lines with vessels under construction was to rescue the company – and so a new firm called Masa was formed, largely with investment from Rederi Ab Slite and from the Floridan Carnival Cruise Line, which was also having a series of ships built at Helsinki at that time. Later, of course, Masa Yards was sold on to the Norwegian industrial, engineering and construction conglomerate, Kværner, and it has continued to expand as a successful builder of passenger ships without further financial problems. For Rederi Ab Slite to ensure the timely completion of the *Athena* and, especially, the *Kalypso,* however, further loans were needed over and above the initial costs of the ships and these, in turn, put the operator under financial duress as well.

When the *Athena* was delivered as planned in April 1989, however, celebration was in order and the giant ship was sent on a celebratory cruise to the Myrsten family's home port of Slite on the island of Gotland, where, dressed overall, she was opened to an admiring public. Indeed, Rederi Ab Slite even chartered the motor coaster *Tella*, formerly its pioneering vessel, the *Slite*, especially for the occasion so that she could be compared with the new flagship.

As the *Athena* was intended to make cruises from Stockholm, rather than ferry crossings as per the remainder of the Viking Line fleet, only part of her car deck was kept clear for cruise passengers to park their cars while they sailed. The remainder was fitted out as an amusement park with amusement arcades, fairground rides and even an aircraft flight simulator. Initially, this remarkable facility was known as 'Röda Lund', a play on 'Gröna Lund', the name of Stockholm's famous and long-established amusement park.

From the outset, the *Athena* was a big success – which was just as well as Rederi Ab Slite had large mortgage repayments to make.

When the rival Silja consortium ordered the *Silja Serenade* and the *Silja Symphony*, however, the Viking Line consortium had understandably felt obliged to outdo its rivals by ordering an even bigger ship. It was Rederi Ab Slite which did the honours. The company's owner, Carl Bertil Myrsten, was so pleased with the success of his existing fleet, which appeared to be making more than sufficient profits to pay off the company's debt, that he ordered a further super ferry, to be named the *Europa*, for the Stockholm-Helsinki route.

An agreement was reached with Meyer Werft at Papenburg in Germany and so confident were both Myrsten and the Swedish bankers who financed the project that they failed to take out any insurance against losses through any sudden increase in the value of the deutschmark against the Swedish kroner. Unfortunately, that is exactly what happened as, by 1990, Sweden, and indeed much of Western Europe, was gripped by recession. The Swedish government devalued the kroner by 10% in 1992, making the *Europa* dearer by no less than 400 million kroner. Unable to finance the completion of its intended new flagship, Rederi Ab Slite was forced to declare bankruptcy and its business was liquidated to pay back the remaining bank loans. Meanwhile, its splendid ships were withdrawn and sold abroad. Both the *Athena* and *Kalypso* went to an Isle of Man-based company backed by the Singaporean Lim casino dynasty, to continue cruising, but from Singapore – significantly itself an area of economic growth. At this point, Rederi Ab Slite's patriarch, Carl Bertil Myrsten, withdrew from the shipping trade. In fact, the *Athena* became the pioneer ship of Star Cruises, now South East Asia's largest cruise line and, as the *Lankapuri Star Aquarius*, she became highly popular on short cruises from Singapore and Hong Kong. Once Star Cruises acquired purpose built tonnage, the *Lankapuri Star Aquarius* was put up for sale.

DFDS purchased her in 2002 and she set sail for Aalborg with a DFDS crew as the *Aquarius*. Shortly after berthing, a mountain of her existing fixtures and fittings began to accumulate on the quayside adjacent in Aalborg's Greenland harbour. By the time she left for Hamburg for dry-docking and a technical overhaul, this mountain had reached her upper decks – crushed up Dampa ceiling by the ton, furniture, cabling, mattresses, carpets, lifesaving equipment. Clearly DFDS intended that she should emerge almost as an entirely new ship. She later returned to Aalborg, where Dansk Interior A/S set about fitting her out with new interiors designed by Figura for her new role as DFDS flagship on the Copenhagen-Oslo route. Having been a Baltic party cruiser, then an Asian casino ship, she would actually be entering service as a car ferry for the first time under DFDS ownership!

About a year later, she loomed over me at the quayside in Copenhagen and I was at last about to sample her delights for myself. As with Crown, DFDS had to create a new entrance hall on one of the ship's lower decks. First off, I inspected the luxury Commodore Class cabins on the topmost deck. These are unaltered from the ship's previous Star Cruises career and those located forward are themed with a variety of Oriental styles. There are tiger skin rugs, Japanese screens, gilded elephants and all kinds of other faux-antique exotica scattered around these most commodious of suites. Those who have been out East will recognise the look from Hotels and other leisure environments in the Pearl Delta cities. Some suites even had large private sundecks with their own outdoor Jacuzzis! Surely the *Pearl of Scandinavia* is the world's only overnight vessel to be so

Explorers Resturant - Crown of Scandinavia *(DFDS)*

Pearl of Scandinavia *(DFDS)*

equipped. The more conventional staterooms further aft date from the Viking Line era and feature windows angled forward and aft.

One deck above is the conference suite, which has numerous rooms and even a large auditorium amidships, which can be subdivided. There is an internet café and a bar for conference guests on this level.

Moving down to the main public room decks, of which there are two, we commenced our tour at the forward end of the ship. Here, on decks 7 and 8, there is a double-deck restaurant facility with the à la carte restaurant located above the buffet restaurant. The most striking feature of the space is the vast expanse of slanted glazing at the front, with a void behind. This design is clearly developed from the four ships built by Wartsila for Silja line in the early-mid 1980s (*Silvia Regina*, *Finlandia*, *Svea* and *Wellamo*).

Aft of the restaurants is an imposing two-level palm court with palm trees and a fine collection of builders' models of previous DFDS 'Oslo boats'. In fact, there are ship models in display cases for much of the length of the vessel's starboard upper arcade on deck 8. Adjacent to this, there is an Oriental/Pacific Rim speciality à la carte restaurant called the Marco Polo, which has a ceramic floor and a wrap-around wall mural. Behind this is a cigar lounge called the Captain's Bar. For sure, DFDS knows how to enjoy the finer things in life. This cosy space consists of large buttoned leather armchairs and a bar with a large, glass humidor displaying a fine range of very large and expensive cigars. The Columbus Club – the main live entertainment space – is located fully aft and has windows overhanging the sides.

The sun decks are magnificent spaces with wooden steamer chairs and even some teak planking to be enjoyed. There is plenty of sheltered deck space atop the ship, protected by glass screens. Aft, there is an outdoor swimming pool, a children's paddling pool and two Jacuzzis. There are also several deck bars and plenty of attractive corners to sit and enjoy the passing scenery. Of course, the *Pearl of Scandinavia* also has an extensive indoor pool, sauna and fitness area forward on her boat deck.

The *Pearl of Scandinavia* is undoubtedly a magnificent flagship for the Oslo route. She looks and feels like a true cruise liner, rather than a cruise ferry (although nowadays the two genres are often close and, at times, indistinguishable).

Bruce Peter

Marco Polo Resturant - Pearl of Scandinavia *(DFDS)*

dfds tor line

re-equips for the millennium

BACKGROUND

When Tor Line started operations in 1966, its initial success and ultimate vanquishing of all other operators between the Sweden and the UK and Sweden and Benelux was based on three major principles – faster ships which necessitated only one night at sea per crossing, ships which could take passengers, cars and freight and high utilisation of ships with fast turn-rounds. However, such was the success of the service and the growth of business that before long it was necessary to consider dedicated freight vessels to supplement the multi-purpose *Tor Anglia* and *Tor Hollandia*. After initially chartering vessels, the first custom built vessels – the 2,540t *Tor Mercia* and *Tor Scandia* were delivered in 1969. Then in 1971 the *Tor Gothia* was delivered in 1971 from the Framnæs Mekaniske Verksted shipyard in Sandefjord, Norway. The ship was designed to take trailers and trade cars on her capacious vehicle decks and the open weather deck could either take more trailers or be crane loaded with containers. This proved to be the first of a series of ten vessels of which six – the *Tor Gothia* (1971), *Tor Belgia* (1972), *Tor Nerlandia* (1972), *Tor Dania* (1973), *Tor Finlandia* (1974) and *Tor Caledonia* (1977) – entered service with Tor Line. All were lengthened from 135.8m to 163.5m in 1977 except the *Tor Caledonia* which was built to 162.7m and lengthened to 188.6m in 1990. The remaining four vessels went to other operators although the *Fichtelberg* of DSR (1976) was initially delivered to Tor Line as the first *Tor Caledonia* and was immediately sold. The *Melbourne Trader* served, as the *Stena Gothica*, for many years on the joint Tor Line/Stena Line Harwich – Gothenburg route and was lengthened at the same time as the *Tor Caledonia*.

These versatile vessels – and other second-hand acquisitions – enabled Tor Line to expand its North Sea services and to establish routes which did not serve Sweden such as Rotterdam – Immingham. However, strong competition and the need to charge competitive rates meant that the company was never able to make the profits its growth would have suggested and in 1981 Tor Line was acquired by the Danish DFDS Group. By that time Tor Line freight operations were a joint venture with Bostroms AB of Sweden (owners of Swedish Lloyd), trading as Tor Lloyd but DFDS bought out both parties and the Tor Line name was revived.

Although Tor Line's passenger operations were soon integrated into those of DFDS, Tor Line's freight services continued to be operated as a separate stand alone subsidiary and DFDS's own smaller freight operation was operated and marketed separately. Whilst all passenger and DFDS ro-ro operations were run from DFDS's HQ in Copenhagen, Tor Line continued to be run from Gothenburg and in many ways the two units operated different policies. For example, Tor Line ships served Immingham whilst DFDS remained loyal to their traditional port of nearby Grimsby until 1995, despite the restrictions imposed by the much smaller lock. In ship purchase policy the two companies differed as DFDS gradually renewed their much smaller fleet with newbuildings – the *Dana Cimbria* in 1986 and the *Dana Futura* in 1996. In contrast, Tor Line did not purchase any new tonnage after 1976, although a number of second hand acquisitions were made and existing ships were enlarged. Tor Line became DFDS Tor Line in 1999 and control moved from Gothenburg to Copenhagen.

NEW INVESTMENT

By the mid-1990s it was clear that what once had been a state-of-the-art fleet was ageing and major new investment was needed. Justifying new investment in any business is never easy and the simple

Tor Suecia at Immingham *(Nick Widdows)*

Tor Magnolia at Immingham *(Nick Widdows)*

argument that something is old and needs replacement cuts little ice with modern day management accountants. So new investment had to be justified on other grounds as well, namely:

- New ships could be bigger ships and carrying more traffic meant more revenue
- Developments in marine technology meant that new ships could be faster which was not only more attractive to customers but meant less ships would be needed to maintain the same number of sailings and, on several routes, ships could operate at regular times throughout the week which pleased customers and made port operations more efficient.
- Developments in engine technology meant that new ships were more efficient, more ecological, less prone to breakdown and could be operated with smaller crews.

These arguments won the day and in 1996 orders were placed with Fincantieri in Italy for three identical ships to be built at their Ancona yard for delivery in 1998. They were to be 195m long, able to cruise at 21 knots and carry 2800 lane metres of freight – the equivalent of about 194 x 13.5m trailers – and would enter service on the busy Gothenburg – Immingham route. Their higher speed would enable a daily service to be operated at the same times each day by three ships rather than the four previously in use. By the standards of the time these were very large ships and although bigger ships have been built subsequently, such ships would not only have been unable to enter the lock at Immingham but could have proved unsuitable on other routes they might operate on during their career.

The basic plan was for two main vehicle decks from bow to stern, uncluttered other than by the two side funnels and ramps from one deck to another, plus a large 'tank top' deck below the main deck. The bridge and accommodation block was raised clear of the upper 'weather' deck and, freed from the need to place it at one end of the vehicle decks as on earlier ro-ros, the opportunity was taken to place it in the traditional place for cargo ships until the 1970s – about 2/3 of the way back from the bow. This layout not only gave the deck crew good sight of both the bow and stern but placed crew and passengers in the most comfortable part of the ship in rough weather and away from engine vibration.

At the turn of 1998, it became clear that progress on the three ships at the Fincantieri Ancona shipyard was not nearly as advanced as was hoped and delivery of the three ships that year was unlikely. The vessels were being built in parallel rather than one after the other and this was leading to slow progress. It was therefore decided that they should be finally completed by other yards.

The first vessel – the *Tor Selandia* – was launched in October 1998 and arrived at the yard of Fredericia Skipsværft A/S, Fredericia in December for final fitting out. She eventually entered service on the Gothenburg – Immingham route the following April. The next vessel – the *Tor Suecia* – was sent to Ørskov Staalskibsværft A/S at Frederikshavn in July 1999 and entered service in October. The final vessel was completed by Fincantieri at their yard at La Spézia and entered service in August 2000.

Once all three ships had been delivered, it was possible to inaugurate a completely new timetable on the route. The four ships previously employed on the route were replaced by three and crossing times slashed from 38 hours to 26. A turn-round time of six hours was allowed at Immingham, with ships arriving at 22.00 and leaving at 04.00 the following morning. An exception was Saturday, when the departure was six hours later to allow shippers extra time to clear out their depots on a Friday night for delivery in Sweden on Monday morning. Having received the last ship some 18 months after the first, the company had had plenty of time to prove that the ships were capable of maintaining these accelerated timings and the new schedule proved highly successful. However, one thing that was clear was that, big as they were, they were still too small to handle the rising traffic levels on the route. Work was therefore progressed on designing a second generation of new ships.

THE SECOND GENERATION

Because of the constraints of the lock at Immingham, new vessels could not be much longer or wider. Extra capacity was thus gained by going upwards – adding an extra deck. The open weather deck of the older design was roofed over – albeit with large openings in the site making it only semi enclosed – with a new weather deck on top, again, the full length of the ship. To increase flexibility and ability to regain time lost through bad weather, the ships would be even faster.

The company placed its first orders for three new ships in 2001. Perhaps not surprisingly in view of the previous late delivery, the order went to a different yard – Flensburger Schiffbau of Flensburg,

Germany. This yard had a reputation for quality and on-time delivery. Initially three ships were ordered for the Gothenburg – Immingham service but this was subsequently increased to five and a sixth ship was ordered in 2004.

Naming policy for the new ships represented a major change from previous practice. Traditionally Tor Line and DFDS Tor Line freighters have had stylised geographical names representing the countries and regions served – eg *Tor Scandia, Tor Hollandia, Tor Flandria.* The new ships were to be named after flowers, reviving the names of DFDS ships of the past – the 1950s and early 1960s built refrigerated cargo ships *Freesia* (1961), *Magnolia* (1963), *Petunia* (1963) and *Primula* (1952) (there was no *Begonia*) which operated mainly between Denmark and UK ports in the days before butter and bacon was carried in refrigerated trailers. Another fifties flower vessel was the *Ficaria*, sister ship to the *Primula*.

The first of the new ships, the *Tor Magnolia*, arrived in October 2003 and replaced the *Tor Britannia* on the Gothenburg – Immingham route. The latter ship moved to the Esbjerg – Immingham service. The second ship – the *Tor Petunia* – arrived in January 2004, allowing the *Tor Suecia* to move to the Esbjerg route and, the following month with two faster ships on the route cutting the journey time to 18 hours, a new regular interval timetable was introduced between Esbjerg and Immingham. With the Gothenburg ships arriving at Immingham at 22.00 and leaving at 04.00 and the Esbjerg ships arriving at 14.30 and leaving at 20.30, the arrivals and departures of ships on the two busiest routes were well spaced across the day. Later in 2004 the *Tor Suecia* received Danish registry and a largely Danish crew.

Next to arrive was the *Tor Primula*, which entered service in May 2004, displacing the last of the first generation ships, the *Tor Selandia*, which moved to the Gothenburg – Ghent route. The delivery of the *Tor Begonia* in September resulted in the first move of a second generation ship when the *Tor Primula* joined her older cousin on the Ghent route. Finally, in January 2005 the *Tor Freesia* arrived, enabling the *Tor Magnolia* to replace the *Tor Selandia* on the Belgian service. This four ship route thus has two of the latest ships which run alternately with the 1981 built *Tor Flandria* and *Tor Scandia*. These ships have similar capacities to the new ships but are a little slower, with a service speed of 19.7 knots. This has made it possible to reduce the sailing time from December 2004, when the arrival time in both Ghent and Gothenburg was changed to 14.00 instead of the previous arrival time of 20.00.

The *Tor Selandia* then took over the Gothenburg – Harwich service, replacing two older, smaller and slower ships (which initially inaugurated a new Immingham – Zeebrugge service). With her greater speed she can undertake two round trips per week compared with the one and a half of the older ships and, with her additional capacity, her two sailings almost equates to the three operated previously. The sixth ship – to be called *Tor Ficaria* – will be delivered in 2006 and there are options on a seventh.

THE SHIPS

Accommodation is provided to a high standard, with both officers and crew enjoying spacious single cabins on the two decks below the bridge. Messing facilities are provided on the lower on these decks with a central galley serving a crew dining and lounge area on the port side and a similar but slightly smaller passenger lounge and dining area on the starboard. Drivers and other passengers are accommodated in ten cabins on the earlier series and twelve on the second series, some double, some single. This enables most or, in the case of the new ships, all passengers to be able to enjoy a single cabin should they want.

The earlier ships have two Sulzer engines generating 10,800 kw of power driving two Kamewa propellers and enabling the ships to cruise at over 21 knots. The propellers are supplements by two 1,100 kw bow thrusters. The later ships have just one engine (and this a single funnel) – a powerful MAN B&W unit developing 20,070 kW. This is linked to a single propeller but as well as the two bow thrusters of the older ships they have two 880 kW stern thrusters. This compensates for the single propeller and rudder and greater windage of the taller ships. Although the newer ships have slightly less main engine power, all onboard electricity is generated by auxiliary engines and the absence of a shaft generator for this means that all power generated goes to propulsion. Thus they are capable of higher speeds.

Both classes of vessel have large state-of-the-art bridges with two comfortable 'drivers' seats surrounded by an array of hi-tech displays, making full use of radar, GPS and AIS ship identification.

Bridge wings are of course fully enclosed and front windows are heated to ensure good visibility in the most wintry conditions.

The business end of any ro-ro is of course the vehicle accommodation. The older ships boast 2,280 lane metres on three levels – a full length main deck, a large 'tank top' below the main deck and a weather deck above, again full length. In the earlier ships a retractable car deck is located in the tank top whereas in the new vessels it is at the rear of the main deck. Main deck capacity in the newer ships is lower because of the extra ramp provided to access the new weather deck directly. Both types of ship are able to accept DFDS Tor Line's patented lashing system for trailers whereby a 'C' shaped metal beam is fixed to the front end of the unit before loading and slots into the deck, obviating the need for fixing chains as on older ships.

Although each ship of the two series was built identically to her sisters, over time each ship acquires her own character and differences occur. For example, following transfer to the Danish flag, the *Tor Suecia* has acquired an English pub style bar in the staff mess room and the sauna – beloved of the Swedes – now sees less use! One thing that has not changed is common messing between officers and crew, although the captain can, if he wishes, take meals in his spacious day room. The following table outlines the main differences:

	TOR SELANDIA CLASS	'FLOWER' CLASS
Length oa	197.5m	199.8m
Beam	26.0m	26.5m
Draft	7.5m	7.35m
Gross Tonnage	24,196	32,289
Net Tonnage	7,258	9,686
Passenger Capacity	12	12
Passenger cabins	10	12
Lane metres – weather deck	1,205	1,272
Lane metres – upper deck	n/a	1,137
Lane metres – main deck	1,210	1,046
Lane metres – tanktop	406	376
Total	2,800	3,831
Engine power	2 x 10,800 kW	1 x 20,070 kW
Propellers	2 x 5.3 m	1 x 6.3 m
Bow thruster power	2 x 1,100 kW	2 x 1,100 kW
Stern thruster power	n/a	2 x880 kW
Service speed	21.1 k	22.5 k

THE FUTURE

With options on more vessels and proven success of the existing fleet it seems highly unlikely that the sixth, as yet un-named, vessel will be the last. Other routes could benefit form the advantages the new ships can bring. For example, if the Immingham – Cuxhaven route – prime route now for the import of BMW cars – could be shortened from its current 22 hours a regular headway could be provided and ships could operate an additional weekly trip. The Humber – Maas route is highly competitive with no less than four operators and both Stena Line and Cobelfret Ferries have new tonnage on order. The building of new river berths outside the existing dock system at Immingham will remove the constraints of the lock on the size of vessels than can operate, although new ship are unlikely to be that much bigger as much larger ships operating on a reduced headway are less attractive to customers and less flexible in a fleet with several routes. In July 2005, DFDS Tor Line take over as main importer of products for the Swedish Stora Paper Group and, as well as bringing increasing traffic through Immingham, a new Gothenburg – Tilbury service will be established, probably replacing the existing Gothenburg – Harwich service and requiring additional tonnage.

Much is unknown at present – but it seems certain that DFDS Tor Line will continue to modernise their fleet to fit themselves for the increasingly competitive ferry market of the Twenty-first Century.

Nick Widdows

the **oldest** shipping company in the world

steam packet reaches 175

The new wooden paddle steamer *Mona's Isle* arrived in Douglas in the middle of August 1830 – less than 8 months after the first meeting to establish a locally controlled steam packet company. She then opened the Liverpool service in the middle of August – a route which the Isle of Man Steam Packet Company still runs 175 years later. In 1832 she was joined by the *Mona* enabling a Whitehaven – Douglas – Dublin service to start, as well as excursions to Garlieston and Kircudbright. The *Queen of the Isle* was built in 1834 – allowing a daily summer service (Sundays excepted) to be run to Liverpool. The final wooden paddle steamer *King Orry* replaced the *Mona* in 1842, and limited services to Fleetwood started – some run jointly with Kemp and Co.

The first iron ship, the *Ben-my-Chree* (1845), used the engines from the *Queen of the Isle*. A weekly Liverpool – Ramsey service started in 1848. Between 1845 and 1876 the company built 11 iron paddle steamers. The *Ben-my-Chree* [2] of 1875 was built at Barrow with two funnels but owing to her slow speed was re-boilered and given four funnels in 1884.

Summer services to Dublin and/or Kingstown (Dun Laoghaire) ran from 1846 onwards, as well as occasional excursion sailings to Garlieston and Kirkcudbright. The Dublin – Douglas – Whitehaven service closed in 1847, but Ramsey – Whitehaven services ran weekly in the 1860s. Between 1866 and 1879 there was a weekly service to Garlieston, with connecting trains on to Glasgow in 1878/9. Regular Fleetwood summer services started in 1876.

The company's last two iron steamers were the single screw *Mona* [2] of 1878 and the twin screw *Fenella* built in 1881. The paddle steamer *Mona's Isle* [2] of 1860 was rebuilt in 1883 as a twin screw steamer and re-named *Ellan Vannin*, but was lost with all hands in December 1909. The first steel ship was the paddle steamer *Mona's Isle* [3] of 1882. With a top speed of 18 knots she carried 1561 passengers.

In 1885 the fleet totalled 10 ships, business was booming and the Steam Packet became a limited liability company. The 19 knot *Mona's Queen* [2], joined the fleet, their last ship with oscillating engines. She remained in service until 1929 and was the company's last paddle steamer. Also in 1885 Belfast sailings started, after the Whitehaven SPCo service closed, and in 1889 Peel – Belfast services were taken over and ran daily in summer until 1901 when lack of dredging at Peel required a return to Douglas. From 1877 to 1880 there were weekly services to Greenore, which then continued to Warrenpoint until 1883.

In 1883 there was a twice weekly summer service to Holyhead, increased to three per week the following year. There was then no service until in 1888 only a six days a week "night mail" service was offered, leaving Douglas at 18.30 and Holyhead at 02.45.

In 1887 the Isle of Man, Liverpool and Manchester Steamship Company (known as Manx Line) built two 21knot paddle steamers, the *Prince of Wales* and the *Queen Victoria* which sailed on the Liverpool route. Both companies suffered losses, and in November 1888 a deal was done whereby the Steam Packet bought both ships and two Manx Line directors joined an enlarged Steam Packet board. The company's next new steamers were the twin screw *Tynwald* [3] (1891), and their last and largest paddle steamer the 21.5 knot *Empress Queen* of 1897.

The company's bankers, Dumbells Bank, collapsed in 1900 with a devastating effect on the Manx economy, and the Steam Packet had to resort to buying older second hand tonnage for the first time. The 1889 built *Dora* was bought in 1901 and re-named the *Douglas* [3] and the *Calais-Douvres*, also 1889, was bought in 1903 from the liquidators of another short lived rival: Liverpool & Douglas Steamers, which had started operations in 1899. The *Calais-Douvres*, the last paddler to join the fleet, was re-named *Mona* [3] but only retained until 1909.

Mona's Isle (1) *(IOMSP Co collection)*

Ben-my Chree (6) *(Miles Cowsill)*

austal

dominating the european fast ferry scene

Based on current public orders, Australian shipbuilder Austal will be the only shipbuilder to deliver large newbuild fast ferries into the European market in 2005, a fact which reflects its current domination of the industry. The three vessels are the 127 metre trimaran *Benchigijua Express* which departed Fremantle for the Canary Islands in early April, the 85 metre catamaran The *Highspeed 5* which is scheduled to take up service in the Greek islands trades for the busy summer season and then a 67 metre catamaran for Virtu Ferries of Malta.

Austal has long attributed its success to its willingness, and ability, to custom design and build vessels for specific customer requirements and even this selection of three vessels demonstrates that clearly. While the trimaran is a very large, and very obvious example of custom design, the level of customisation in the catamarans is also worthy of note. For example, the six engine configuration on the *Auto Express 67* for Virtu Ferries, believed to be an industry first, was developed so that the owneŕs requirements for propulsion redundancy, capacity and speed could be met in the optimal way with the available engine technology. Items such as the vehicle and passenger ramps, not to mention the anchor-handling arrangement, on the *Highspeed 5* also represent a significant level of detail design to meet the needs of the Mediterranean style mooring used in Greece. The trimaran is regarded as a major breakthrough, developed over many years as a result of customer Fred. Olsen, SA's identification of a need for fast ships with improved motion characteristics. Going forward, Austal believes that trimarans will prove attractive to many operators particularly those whose operations face challenging sea conditions.

BENCHIGIJUA EXPRESS

Austal describes the Auto Express 127 trimaran as quite simply the most significant vessel to arrive on the fast ferry stage and says it is set to allow fast sea transportation to improve and open up new markets beyond the ability of existing fast ferry design for both commercial and military operators. When Austal and European ferry operator Fred. Olsen, S.A. signed a contract in Santa Cruz de Tenerife in May 2003 to build a second high speed vehicle/passenger ferry, it was the signal to start constructing not just another ferry but also the end to an exhaustive design, research and development programme with origins dating back as far as early 2000, to create a new hullform which would provide a quantum improvement in the performance and operability of large fast ferries.

Fred. Olsen, S.A. pioneered the use of large high-speed ferries in the Canary Islands and currently carries almost three million passengers, half a million cars and a quarter of a million cargo vehicles per year. Since 1999 it has been operating large fast catamarans and while happy with the results, the company identified some limitations in terms of capacity and especially passenger comfort when operating in rough seas. This led to the conclusion that further research was needed in order to develop a new concept for high-speed vessels, combining the softer roll of monohulls with the low resistance, very good stability and carrying capacity of catamarans.

With these objectives in mind, Fred. Olsen, S.A. and Austal cooperated on an extensive programme of research, tank testing and analysis, firstly to develop a new design and then ensure it would meet Fred. Olsen, S.Aís requirements in an efficient and cost-effective manner. The result is correctly described as a slender stabilised monohull, but more commonly referred to as a trimaran.

Austal Chairman, Mr John Rothwell, is confident the superior seakeeping performance of the trimaran will provide Fred. Olsen, S.A.'s passengers with significantly enhanced levels of comfort compared to the companyís existing fast ferries and is also expected to result in noticeably higher

Highspeed 5 *(Austal)*

Benchigijua Express *(Austal)*

AUSTAL
FRED. OLSEN Express
BENCHIJIGUA EXPRESS
THE FUTURE
HAS ARRIVED
BIG SHIP COMFORT
IN A FAST FERRY
AUSTAL 100 Clarence Beach Road, Henderson, Western Australia 6166
Tel +61 8 9410 1111 Fax +61 8 9410 2564 Email: marketing@austal.com
www.austal.com

levels of operability. "Studies based on actual sea conditions found in the Canary Islands enabled Austal to accurately model the vesselís performance on each leg of the proposed route. The final report showed the trimaran would offer a 26% improvement in operability compared to existing hullforms", he explained.

Fred. Olsen, S.A. Chairman, Mr. Fred Olsen Jnr. said that through the close cooperation with Austal a highly versatile vessel with many improvements over his current fast ship fleet has been developed. "The characteristics of this new vessel, with a length of 126.7 metres and beam of 30.4 metres, will improve overall efficiency in terms of passenger capacity, deadweight and freight lane metres by more than 35%. At the same time passenger comfort will increase by 25% to 40% depending on the routes we operate", Mr. Olsen said.

"This trimaran should, for us, be the solution for many years to come and could very well set the standard for a new generation of large fast ferries. We believe our customers deserve the best"

During sea trials with operating ride control *Benchijigua Express* achieved a speed of 40.4 knots whilst carrying deadweight of 500 tonnes. With a capacity to carry 1,350 passengers and 341 cars the ferry will operate between Los Christianos in the south of Tenerife and the islands of La Gomera and La Palma.

Arranged in two separate engine rooms in the trimaranís central hull are four MTU 20V 8000 diesel engines, each rated to 8,200kW which will be upgraded to 9,100kW during the first quarter of 2006. Those in the aft engine room each drive a Kamewa 125 SII steerable waterjet from Rolls-Royce while the two forward engines deliver their combined power to a Kamewa 180 BII booster waterjet. Each of the three drivelines features Renk transmissions. "Even though the trimaran is very much larger in terms of both length and capacity, the challenge for Austal has been to deliver manoeuvring characteristics equivalent or better than Fred. Olsen S.A.'s existing vessels", said Mr. James Bennett, Austal's Technical Manager.

This has been achieved by fitting two Ulstein Aquamaster UL601 azimuthing bow thrusters supplied by Rolls-Royce. The ability to synchronise the thruster and waterjet control systems gives the vessel's Captain excellent control to ensure fast, efficient and safe operation in port. Once in open water the electrically driven thrusters are retracted into the hull to reduce drag.

Initial full power sea trials in conditions of up to four metres significant wave height and 45 knots wind speed confirmed Austalís confidence in the 127 metre trimaran fast ferry. Whilst the vessel met the key contract performance requirements it, most importantly, also demonstrated the trimaran's ability to exceed the operability and comfort levels of existing large catamaran designs at the upper end of the weather spectrum.

Austal is now working closely with Flag States to achieve operability certification for the Auto Express 127 trimaran in 4 metres, or more, significant wave height.

HIGHSPEED 5

The Auto Express 85 fast car ferry *Highspeed 5* was launched at the Austal shipyard on March 20. The 85 metre aluminium catamaran is the fifth fast ferry the shipbuilder has built for leading Greek ferry operator Hellas Flying Dolphins, which has recently re-branded its operations under the new name Hellenic Seaways.

Scheduled for delivery in May, the custom designed ferry will enter service following its delivery voyage from Australia to Greece. Its operations will include a new fast ferry service linking Piraeus with Chania on the island of Crete; one of the most important routes in the Greek Islands trades. The Austal Auto Express 92 *Highspeed 4* (delivered in July 2000) inaugurated that service earlier this year.

When complete, *Highspeed 5* will be able to carry 810 passengers and 154 cars and will be capable of 39 knots when loaded. The catamaranís propulsion consists of four Caterpillar 3618 diesels each driving a Rolls-Royce Kamewa waterjet via a Reintjes gearbox.

Austal Sales and Product Development Manager Mr Glenn Williams says the contract rewarded the marketing effort the company has dedicated to the Greek market over many years. "We have been

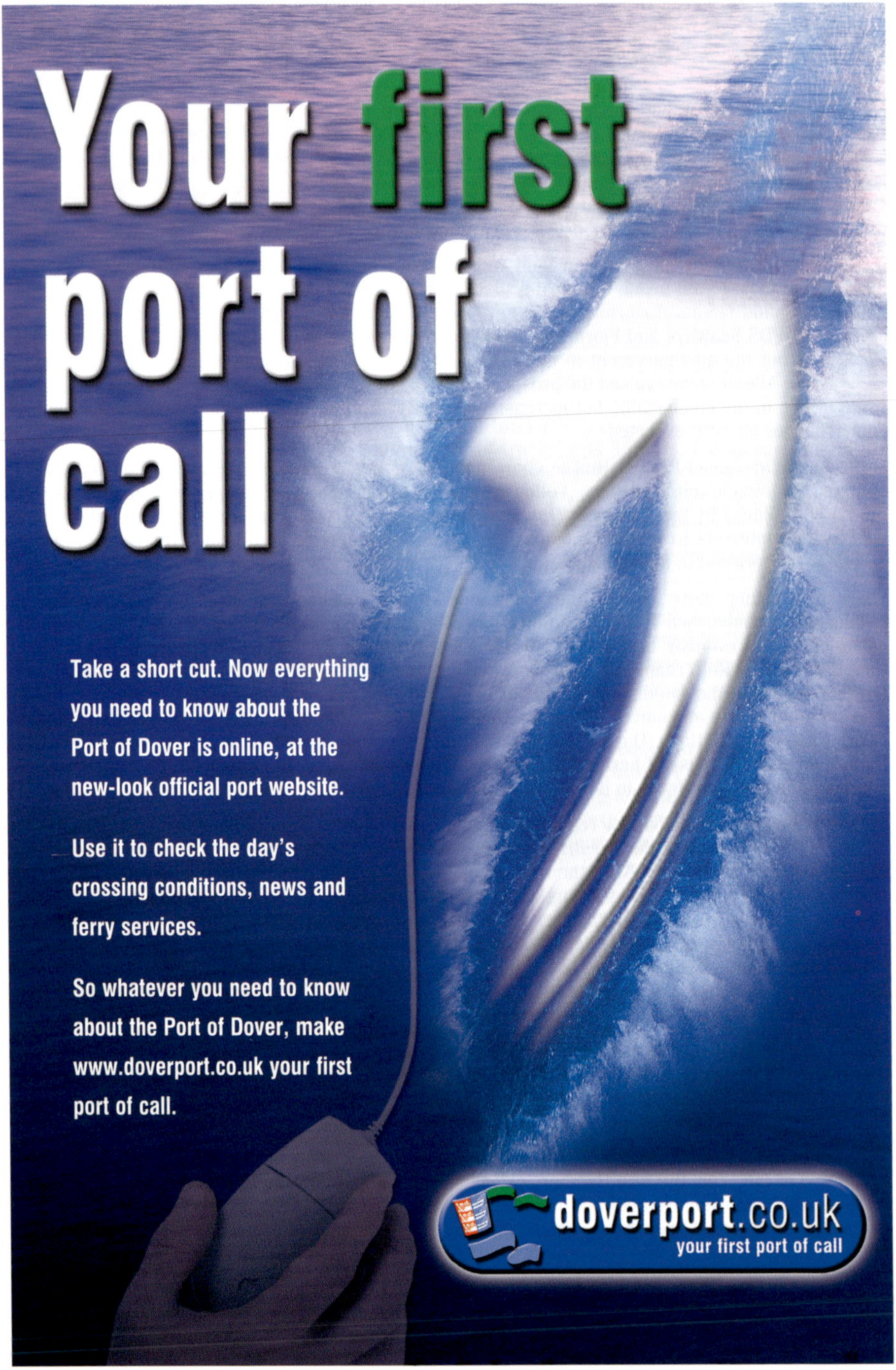
Your first port of call
Take a short cut. Now everything you need to know about the Port of Dover is online, at the new-look official port website.
Use it to check the day's crossing conditions, news and ferry services.
So whatever you need to know about the Port of Dover, make www.doverport.co.uk your first port of call.
doverport.co.uk
your first port of call

The Haven ports of Harwich, Felixstowe and Ipswich continued much as before during 2004 with less change than during 2003. However, during the summer DFDS Tor Line deployed the Lisco vessel *Palanga* on the Harwich – Esbjerg route to relieve pressure on the ro-pax *Dana Sirena* during the busiest time for passenger traffic. The Gothenburg – Harwich service ceased to be a joint DFDS Tor Line/Stena Line operation and the *Stena Gothica* was replaced by the *Tor Neringa*. Two of the three weekly Harwich – Gothenburg sailings began operating via Rotterdam in May as the 'Maas bridge' but this ended in December and early in 2005 the service became a single ship operation with the *Tor Selandia* performing two round trips per week.

THAMES AND MEDWAY

2004 proved rather less newsworthy than 2003 when charters of vessels for the Iraq war caused major alterations to ship working. In the spring Cobelfret Ferries swapped the Luxembourg registered *Melusine* for the chartered *Louise Russ* – still in RORO Express colours on the Purfleet – Rotterdam route. Apart from that the operator had a quiet year.

On the other side of the river, Dart Line chartered the AWSR vessel *Eddystone* to cover for the *Dart 9's* dry docking in November. The *Dart 2*, used for most of the year on the Dunkerque service, was chartered to Channel Freight Ferries, also in November and replaced by the chartered *Breant*. This vessel seemed unable to cope with the strong Thames tides and regularly ran very late. Shortly afterwards, Dart Line announced that the Dunkerque service would cease after 10th December and when this happened the charter of the *Breant* was terminated. The *Dart 2* returned in early January 2005 and was placed on the Dartford – Zeebrugge route to cover for the dry docked *Dart 8*.

The Gravesend – Tilbury ferry *Duchess M* was out of action for several days in the autumn after being struck by *HMS Grimsby* undertaking an exercise in the river. Her owners' other vessel, the excursion ship *Princess Pocahontas* was also slightly damaged but was able to continue sailing and took over the ferry's role while repairs were carried out.

EASTERN CHANNEL

TransEuropa Ferries introduced their second ex Townsend-Thoresen/P&O freighter the *Begonia* (ex *European Pathfinder*) early in the New Year. In July the much-awaited passenger service was launched, using the *Oleander* and *Larkspur* to make four sailings a day available to passengers with cars. Subsequently the *Eurovoyager* was also used in passenger mode when working the one of the same rosters.

In January, P&O withdrew the *European Seaway* from the Dover-Calais route and began operating the *Pride of Burgundy* in freight only mode. The reasoning behind this – rather than simply withdraw a passenger ship – was presumably so she could be switched to full passenger mode if necessary and this did in fact happen for a brief period during the summer when another ship was out of action. At the same time, P&O reduced facilities on their night time sailings and ceased to carry foot passengers, branding the service 'Lite-Night'. Although the company considered chartering out the *European Seaway* or moving her to another route, this did not happen and she spent the year laid up at Falmouth and Birkenhead.

These changes were a precursor to more announced in September. This involved the disposal of the *Pride of Provence* but the reinstatement of the *European Seaway* and the *Pride of Burgundy's* reverting to full passenger mode. The *Pride of Aquitaine* was to return to her owners at the end of 2005 when her charter finished. On the face of it, the impact did not appear to be too great but behind the rather vague talk of an 'improved service package' on-board lay over 600 job cuts at Dover and the ending of the *Pride of Aquitaine's* service in May 2005. The announcement was followed by protracted discussions with the trades unions but eventually agreement was reached without recourse to industrial action. The *Pride of Provence* duly went out of service in December, being sold to GA Ferries of Greece, and the *European Seaway* came back in January 2005, followed by the *Pride of Burgundy* as a passenger ship once more.

SpeedFerries, who had planned to start operations in 2003, eventually launched their Dover – Boulogne service in May 2004 using, not the Mols-Linien 91m InCat *Mads Mols* as originally planned but the 86m craft which had at one time operated for the Australian Navy as *HMAS Jervis Bay*,

Oostende
roll-on, roll-off in Oostende,
it's fast, it's easy
The port of Oostende in Belgium offers you a long-established know-how in handling ro-ro freight. Its up-to-date infrastructure includes versatile berthing facilities and well-situated parking areas. Major terminus of European highways, inland waterways, railroads and an international airport make Oostende a true global cargo hub.
Killingholme
Ipswich
Ramsgate
OOSTENDE
Haven
Oostende
www.re-action.be
Port
Oostende
AG PORT OOSTENDE-BELGIUM
Slijkensesteenweg 2 - B-8400 Oostende
Tel. +32(0)59 34 07 11 - Fax +32(0)59 34 07 10
website: www.portofoostende.be - e-mail: info@portofoostende.be

renamed the *SpeedOne*. The company no longer had any connection with Mols-Linien and plans for Norfolkline to provide handling arrangements at Dover were dropped. This meant that the company had no booking office in the Dover Travel Centre so all sales had to be done by phone or Internet. Pricing was very much on the lines of low-cost airlines with return fares in peak periods as low as £60 provided you booked long enough ahead and the fares were still available. The lack of a booking office was probably one of the reasons no foot passengers were conveyed, much to the chagrin of the traders of Boulogne who were looking forward to the restoration of a ferry service to bring them British day-trippers once again. The service was launched with a certain amount of ill feeling between the new company and established operators and there were allegations of 'dirty tricks' against the 'interlopers', and their craft was emblazoned with the slogan 'Fight the pirates'. Eventually an agreement was reached with SeaFrance that they would carry SpeedFerries' customers to and from Calais should the fast ferry not be able to operate.

Hoverspeed operated what is likely to their last season using 78m Incats – which first came to Dover in 1991. Most of the year the 78m *SeaCat Scotland* operated with the 81m craft *Diamant* (renamed the *SeaCat Diamant* in the autumn) but in the peak they brought in the *SeaCat France*, laid up since the end of summer 2002.

SeaFrance continued to operate the *SeaFrance Renoir* throughout the year although she operated with reduced facilities and foot passengers were excluded. They announced that their newbuilding due in 2005 was to be called the *SeaFrance Berlioz*, named after a composer rather than a visual artist.

Norfolkline announced that they had ordered a third new ship from Samsung in Korea. Like her sisters ordered the previous year, she will have slightly less vehicle accommodation than the existing vessels chartered from NorseMerchant Ferries but more passenger accommodation and will be faster, accomplishing the Dover – Dunkerque service in 1hour 50 minutes.

Newhaven operator Transmanche Ferries continued their operations as before although there were a number of incidents of ships grounding in the treacherous approach to Newhaven Harbour. In the autumn, discussions were held with Sea Containers regarding Transmanche taking over the summer only fast ferry service, possibly time chartering the *SuperSeaCat One*. However, these talks finally came to nothing at the route was withdrawn.

WESTERN CHANNEL & SOLENT

P&O started the year in buoyant mood by announcing the operation of a fast ferry service to Caen using the chartered *Max Mols* as the 'Caen Express'. The service was quite successful and the craft, with Danish officers, ran reasonably well. However, in the autumn came the announcement that, as part of their review, P&O were to end all ferry services from Portsmouth to France, both fast and conventional ferries. The two Le Havre ships were to be sub-chartered to Brittany Ferries and the crews would pass to their employ. The *Max Mols* was to return to Mols-Linien (who were quite pleased to get her back due to growing traffic on their own routes) and the *Express* (trading as 'Cherbourg Express') would move to the Larne-Cairnryan/Troon service in 2005. A sub-charter would be found for the *Pride of Cherbourg* and the *Pride of Bilbao* would operate between the UK and Spain only, no longer operating her popular Friday night 'booze cruise' to Cherbourg. It later emerged that the charter from Irish Ferries of the *Pride of Cherbourg* would be transferred to Stena RoRo who would rename her the *Stena Challenger* and the *Pride of Le Havre* and *Pride of Portsmouth* would, respectively, be renamed the *Honfleur* and *Etretat*. However, the latter plans were put on hold in November when the Office of Fair Trading decided that the plans of the two companies constituted a merger and the matter must be referred to the Competition Commission who would not report until May 2005. This created more delay and uncertainty for both staff and customers and increased loses for P&O. It is difficult to see any other sensible solution other than complete closure of the route, which would throw everyone out of work. None the less, rather than do that or challenge the decision in the courts, P&O decided to continue the service until the results of the Commission's deliberations were known. The Cherbourg route closed on 14th January 2005 with the *Pride of Cherbourg* going to Stena Line to operate between Karlskrona and Gdynia whilst the *Stena Baltica* was undergoing a major rebuild. Subsequently, Brittany Ferries announced that in view of the imposed delay caused by the Competition Commission referral, they were pulling out of the agreement, so unless something dramatic happens this summer, the route will close this autumn.

SeaFrance Manet *(Miles Cowsill)*

Pride of Calais *(Miles Cowsill)*

Brittany Ferries, launched a Portsmouth – Cherbourg service, albeit with less sailings than originally advertised. Towards the end of the year it was announced that, in 2005, a conventional ship would operate to Cherbourg until March and then a fast ferry – the *Normandie Express* – would take over. Although not shown in the timetable – because it would have presumed the decision of the OFT – it was intended that throughout the year, the former P&O Le Havre ship arriving overnight would operate a round trip to Cherbourg rather than Le Havre before operating the next night's service to Le Havre. The eventual OFT decision meant a rethink and it was subsequently announced that the backup conventional sailing to Cherbourg would initially be provided by the Cherbourg – Poole vessel the *Barfleur* undertaking a day sailing to Portsmouth rather than Poole.

Portsmouth Harbour Ferry Company – operators of the Portsmouth – Gosport link – were taken over by the Falkland Islands Company. A sister vessel for the 2000 built *Spirit of Gosport*, the *Spirit of Portsmouth*, was ordered during 2004. Meanwhile, plans for the South Hants Light Rail system to provide a link in tunnel between the two ports were put on indefinite hold when the Department of Transport said that the plans for the network were too expensive and must be re-evaluated.

Channel ferries returned to Southampton in 2004 in the form of Channel Freight Ferries, a company owned by shipbrokers Clarkson's and managed by Simon Taylor who had previously launched Dart Line and RORO Express. The service ran from Southampton to Radicatel – a new installation on the Seine, near to Le Havre but part of the Port of Rouen and initially an overnight service in each direction was operated. The ships used were the *Bazias 1* (renamed the *CFF Seine*) and the *Tango* (renamed the *CFF Solent*) – both owned by Octogon Shipping of Romania. Traffic grew during the year until, by the autumn, it was decided to offer a twice-daily service. However, the *Tango* with her internal lift rather than ramps, was deemed unable to meet the more stringent turn-round times such a service would involve and her charter was ended. However, the *CFF Seine* however was purchased by the company.

The *Dart 2* – a sister vessel of the *CFF Seine* was chartered in November and it was hoped that with two Bazias class vessels and their excellent loading arrangements a twice daily service could then start. However, this did not prove possible and the charter ended in early January 2005. The faster Estonian owned *Lembitu* replaced her and a twice-daily service – three days per week – started at last on 11th January. However, at Easter, the *Lembitu* left and a single ship operation was introduced.

Red Funnel Ferries' *Red Falcon* and *Red Osprey* both went to the Remontowa Shipyard in Gdansk for stretching and the addition of an extra vehicle deck over winter 2003/2004. Their replacement was the former Norwegian fjord ferry *Nordhordland*, renamed the *Bergen Castle*. Offering a much lower level of passenger comfort (her utilitarian passenger accommodation was below the vehicle decks) she was not very popular with customers and was laid up when the second rebuilt ferry returned. When the third 'Raptor Class' went away in December 2004, it was decided that that two remaining ships, now with greater capacity, could handle the winter traffic on their own.

Plymouth's year started with great hope with the arrival of Brittany Ferries' new 27- knot super-ferry, the *Pont-Aven*, able to operate a faster service to Spain and an extra round trip to Roscoff. All went well until 10th August, at the height of the summer peak, when the ship failed completely on the Plymouth linkspan. 1,200 tonnes of seawater had flooded an auxiliary engine room through a faulty cooling valve. She was unable to carry her own traffic and the Roscoff ferry, the *Duc de Normandie*, could not berth either. She was eventually towed to Brest for repairs but, on leaving France, had to put back for further work. She eventually re-entered service from Santander on 24th August but was forced to keep her speed to 18 knots which meant longer crossing times and the loss each week of one round trip from Roscoff to Plymouth. She went for further repairs at the ARNO shipyard at Dunkerque on 22nd November and remained there for just under a month before taking on her winter role between Portsmouth and St Malo.

The *Duc de Normandie* was laid up for sale at the end of the summer period and eventually purchased by TransEuropa Shipping.

Channel Islands operator Emeraude Lines faced major problems at the beginning of the year due to a dispute between the new owners of the company and the owners of their main vessel, *Solidor 5*. Initially the *Solidor 4* (renamed the *Aline* for a charter to Senegal which did not in fact occur) operated their services but later she was replaced by the *Hoverspeed Great Britain*, chartered from Sea Containers. She was initially marketed as the 'Grand Bretagne' but subsequently formally renamed the

Emeraude GB. She operated between Jersey and St Malo only. The company was rebranded as Emeraude Jersey Ferries. A controlling interest in rival Condor Ferries was taken by The Royal Bank of Scotland for £240 million.

French operator Hugo Express, who had taken over Emeraude's passenger only routes, operated both their craft for a full season after a somewhat hesitant start in 2003. Prospects for a new service between Alderney and Guernsey and Alderney and Diellete in France by a new company called it Ferries foundered following a dispute between the operators and the Tyneside builder of the vessel, the *Braye Spirit*.

The end of the year saw the delivery of the first of three new vessels for the busy Torpoint ferry near Plymouth. The *Plym II*, built by Ferguson Shipbuilders in Scotland was towed from her builders as she has no propulsion system, crossing the river by chains. She replaced the 1968 built *Plym*. Replacements for the other two vessels – the *Lyhner* and *Tamar* are being delivered in 2005. They each offer a 50% increase in capacity on the older vessels.

IRISH SEA

After Stena Line returned the *Stena Lynx III* to American Fast Ferries at the end of the 2003 summer season, it was not expected she would return. However, they subsequently decide to purchase her and she resumed service on the Fishguard – Rosslare route in spring 2004.

In March NorseMerchant Ferries restored a partial passenger service on their Birkenhead – Dublin route on sailings operated by the *Brave Merchant*. They announced that in 2005 a full passenger service would be restored by the replacement of the vessels used on the Birkenhead – Belfast service by new ships currently under construction for Levantina Trasporti at their associate shipyard, CN Visentini of Donada, and the transfer of these ships to the Dublin route. The first transfer is likely to take place in July 2005 when the *Brave Merchant's* running mate, the 12 passenger *Lindarosa*, is replaced by the *Liverpool Viking* (previously the *Lagan Viking*). With the *Brave Merchant*, a full passenger service will again be possible. During the year the company chartered the *Leili* (a sister of the *Varbola*) from her Estonian owners to operate additional sailings between Liverpool and Dublin.

The Isle of Man Steam Packet Company continued to operate the seasonal Liverpool – Dublin service they inherited from Sea Containers but at the end of the year they announced it would not resume in 2005. The *SeaCat Isle of Man* was to be retired and all fast ferry services operated by the chartered *SuperSeacat Two*. The Douglas – Heysham ferry *Ben-my-Chree* re-entered service after her winter refit with additional passenger accommodation and her passenger certificate increased in early 2005 following the installation of additional safety equipment. The veteran *Lady of Mann* once again spent her summer operating for Acor Line in the Azores.

Stena Line took over P&O Irish Ferries' Fleetwood – Larne service in March. The three vessels used on the route were taken over and were simply renamed by replacing the word 'European' with 'Stena' (the names *European Leader*, *European Pioneer* and *European Seafarer* which replaced the Pandoro animal names in 1997/8 could have been designed for an ultimate takeover by Stena as they fitted neatly into that company's naming policy for freighters, in all ending with 'er'). P&O's Mostyn – Dublin route ceased on 4th April and both ships used were quickly disposed of – the *European Ambassador* to Stena Line where she became the *Stena Nordica* and the *European Envoy* to Kystlink as the *Envoy*. The 'Ambassador's' weekend Dublin – Cherbourg trip also ceased but the Liverpool – Dublin route, using the former Hull – Rotterdam ro-pax vessels *Norbank* and *Norbay* remained with P&O as required by the Competition Commission.

At Heysham, NorseMerchant Ferries sold the *Merchant Bravery* and *Merchant Brilliant* to foreign interests and chartered them back. They were transferred to the Jamaican flag.

IRELAND

The operators of the Lough Foyle Ferry service opened a second route in the spring, this time wholly within the Irish Republic. It was between Buncrana and Rathmullan across Lough Swilly and the vessel used was a former River Weser ferry re-named the *Foyle Rambler*.

P&O's Rosslare – Cherbourg service ended in December as part of the company's service review. It

Jonathan Swift *(Miles Cowsill)*

Hebridean Isles *(CalMac)*

Stena Carisma *(Miles Cowsill)*

Sagafjord *(FERRYInformation)*

Varberg harbour. There were at one times fears that she would capsize but, fortunately, this did not happen. After temporary repairs were effected she was moved to the Cityvarvet shipyard at Gothenburg for a further assessment of damage and hull repairs. She was then moved to Gdansk for full repairs and returned to service at the end of May. No substitute vessel was provided – traffic was diverted to the Gothenburg – Frederikshavn service.

Kyst-Link, who operate a ro-pax service between Langesund and Hirtshals, acquired a new vessels in the spring in the form of the former P&O Irish Sea vessel *European Envoy*, which was renamed the *Envoy*. She replaced the *Boa Vista*, the one time *Normandie Shipper* and *Speedlink Vanguard*. The *Envoy* had similar passenger capacity but almost twice the freight space of the vessel she replaced.

The new Saga Line service between Moss and Skagen started at last in May. The delay was due to securing a passenger certificate for the *Sagafjord*, which had been out of use since withdrawn by Color Line in 2000. However, the service did not prove as successful as had been hoped and was withdrawn at the end on October and the company later went into liquidation.

JetLink, another Norwegian company trying to get into the supposedly lucrative Norway – Denmark market, announced their intentions of opening a fast ferry service between Moss and Frederikshavn. The vessel to be used was the 78m InCat which started life as the *Condor 11* in 1994 and the following year became the *Cat-Link II*. She had been laid up in Portland, Dorset since 2002, when she was renamed the *Elanora*. To be renamed the *JetLink Express*, she would make the Kattegat crossing in four hours. However, due to various delays the service did not start in 2004. The launch is now planned for May 2005.

In January, Stena Line chartered the freight ro-ro *Bore Mari*, to provide additional capacity on the Gothenburg – Travemünde route pending the arrival of new tonnage. Both the existing ships – *Stena Carrier* and *Stena Freighter* – were, in December, sold to Lillbacka Global Transports of Finland to launch a new Travemünde – Turku service with delivery when replacement tonnage was received. The *Stena Freighter* was renamed the *Global Freighter* in January, whilst still in Stena Line service and in February she was replaced by the *Stena Foreteller*, one of the Chinese built Stena 4Runner Mark 2 class. However, the charter of this vessel from sister company Stena RoRo was still only a short term measure. The longer term replacements were two Stena 4Runner Mark 1 class vessels, laid down in Italy in 1998/99 but not completed until 2004. One, the *Stena Freighter*, was purchased from the liquidators of the Italian SEC shipyard by Stena Line, who had her towed to Kraljevica in Croatia for completion. The second – the new *Stena Carrier* – was initially purchased by Enrico Bugazzi Shipmanagement, who renamed her the *Aronte* and, in 2003 towed her to Marina di Carrara for completion. The *Stena Freighter* entered service in March and replaced the *Stena Foreteller*. The *Stena Carrier* entered service in April as the *Stena Carrier II* (she was renamed the *Stena Carrier* in July). She replaced the previous *Stena Carrier*.

DANISH DOMESTIC

Mols-Linien placed the *Max Mols* along side the *Mads Mols* to provide a two vessel service on the Århus – Odden service when the former ship returned from her charter to P&O in November. The Odden – Ebeltoft route remained a single ship operation, although a two ship service was operated over the Christmas period, including Christmas morning!

Scandlines decided that they could operate the Spodsbjerg – Taars service on a commercial basis without seeking subsidy from the Great Belt Bridge so the Danish authorities cancelled plans for a tendering exercise. BornholmsTrafikken ordered two newbuildings – the *Dueodde* and the *Hammerodde*, – to be delivered in April 2005 by the Merwede Shipyard in The Netherlands. Because of the tight deadline set the company by the Danish Government after they awarded them the franchise, BornholmsTrafikken were forced to order ships for which designs already existed and thus chose two ships following closely the design of the IOMPS's *Ben-My-Chree* and Condor Ferries *Commodore Clipper* (Merwede Shipyards were last owners of the now closed van der Giessen-de Noord shipyard which built these ships). The new ships will replace the *Jens Kofoed* and *Povl Anker* on services from Bornholm to Sealand and Sweden although the latter ship will be retained for a seasonal route to Sassnitz. The chartered freighter *Vilja* was withdrawn in the autumn when the passenger service was transferred to the port of Køge, south of Copenhagen. The charter of the fast ferry *Villum Clausen* was extended for a further five years.

SOUTH BALTIC

The fitting of new engines to the Scandlines vessels *Kronprins Frederik* and *Prins Joachim*, originally planned for winter 2003/4 was deferred until the autumn. The *Kronprins Frederik* was fitted with a Kappel propeller which will enable comparisons to be made with her sister's conventional propeller in terms of fuel efficiency. The two ships started their new faster timetable on the Gedser – Rostock route on Monday, 31st January 2005, enabling the *Dronning Margrethe II* to be withdrawn and sold.

Scandlines announced that the Sassnitz – Rønne service would not resume in 2005 for financial reasons and the veteran *Rügen*, of 1972 would be sold

TT-Line ended the fast ferry service between Rostock and Trelleborg in December and the Austal fast craft *TT-Delphin* was laid up.

A fourth ship started operating for Nordö-Link between Travemünde and Malmö in January when the *Finnsailor* was moved down from FinnLink. The company now conveys 50% of freight between Germany and Sweden.

Stena Line took possession of the former P&O Irish Sea Mostyn – Dublin vessel *European Ambassador* in the spring. Renamed the *Stena Nordica*, she replaced the *Stena Traveller* on the Karlskrona – Gdynia service. The *Stena Traveller* was sold to Lisco Baltic Service and renamed the *Lisco Patria*. She was placed on the Klaìpeda – Karlshamn route. Stena RoRo took over the charter of the P&O Ferries' *Pride of Cherbourg* in January 2005. She was renamed the *Stena Challenger* and initially operated for Stena Line between Karlskrona and Gdynia during the rebuilding of the *Stena Baltica* between February and April. The *Stena Baltica* has had two half decks of cabins converted to a second trailer deck with additional cabins built higher up and a general modernisation of passenger facilities.

DFDS owned Latlines, which operates a ro-pax service between Lübeck and Riga, was renamed DFDS Tor Line SIA and effectively became a marketing organisation with ships operated by DFDS owned Lisco Baltic Service. In January, the *Mermaid II* was replaced on the route by the *Kaunas* of Lisco. The *Mermaid II*, in turn, replaced the 12 trailer ferry *Siauliai* on Lisco's Karlshamn – Klaìpeda route. The *Lisco Patria* (see above) joined the route in the spring, replacing the *Palanga*.

DFDS Tor Line's Fredericia – Copenhagen – Klaìpeda freight service was reduced to a single ship operation, using, initially, the *Tor Neringa* and later the chartered *Stena Shipper*. Latvian Vent-Lines closed their Travemünde – Ventspils service in July 2004 following the sale of the chartered *Kaptan Burhanettin Isim* to Polsteam (Polish Steamship Company). DFDS later revived the service using Lisco's *Palanga*, displaced from the Klaìpeda – Karlshamn route in May by the *Lisco Patria* and between June and August operating between Esbjerg and Harwich as back-up to the *Dana Sirena*. At the end of the year, the *Palanga* was chartered to Trasmediterranea for service between Spain and the Canary Islands. The *Kaunas* was transferred from the Lübeck – Riga route to the Klaìpeda – Karlshamn route and the chartered *Mermaid II* moved from that route to operate on both the Lübeck – Riga and Lübeck – Ventspils routes.

DFDS Tor Line started their twice weekly PolBridge Gothenburg – Copenhagen – Gdansk service on 22nd September. Although it was originally planned to use the *Tor Cimbria* on this route, she had been chartered out so, instead, the *Tor Belgia* launched the service. Because traffic levels were lower than expected, she was replaced in November by the chartered *Tango*, previously used by Channel Freight Ferries. The service ceased in February 2005.

The *Kaptan Burhanettin Isim* – renamed the *Gryf* – entered service with Unity Line on 3rd January 2005 after a major rebuild. Although expected to replace one of the two older train/lorry ferries, she actually operated as an extra vessel, giving, with the passenger vessel *Polonia*, a four ship service.

Rederi AB Gotland ordered a 122 metre fast monohull from Fincantieri. The new vessel, to be delivered early in 2006, will operate for subsidiary Destination Gotland and have a steel hull and aluminium superstructure. She will have a capacity for 800 passengers and over 160 cars. The company's existing French built fast ferry, the *Gotlandia*, also has a steel hull, which was specified in order to cope with winter ice, although in the event the Swedish Government has declined to provide sufficient subsidy for winter operations.

Finnlines subsidiaries Nordö-Link (Malmö – Travemünde) and FinnLink (Kapellskär – Naantali) swapped

Gotlandia *(Miles Cowsill)*

Stena Carrier *(FERRYInformation)*

ships early in 2005. The *Finnclipper* moved to the southern route whilst the Indonesian built *Finnarrow* moved north.

Polferries purchased GA Ferries' *Alkmini A* (formerly P&O Ferries' *PO Canterbury*) and renamed her the *Wawel*. In February 2005, following a major rebuild to install 450 passenger berths she replaced both the *Silesia* and the charted *Kahleberg* (which was withdrawn in late 2004) on the Swinoujscie – Ystad service.

CROSS BALTIC

In February, Finnlines placed an order for three new ships to replace the four 'Hansa Class' vessels which operate a daily service between Travemünde and Helsinki, with two nights on board. Some 90 passengers are conveyed in considerably comfort. The new ships are being built by Fincantieri of Italy and will convey up to 500 passengers. At 216.0m they will be among the longest short sea ferries ever built and will have 4,200 lane metres for freight. With a service speed of 25 knots they will be able to reduce the journey time.

After a major rebuild, the Silja Line *Finnjet* launched a new Rostock – Tallinn – St Petersburg service in the spring. Designed to carry mainly mini-cruise passengers and freight on the vehicle decks, the new service had a moderately successful summer but, although planned to be a year round service, it was withdrawn for the winter in September, with a promise that it would resume on 28th April 2005. Meanwhile the ship was laid up at Rostock.

Riga Sea Line introduced a second vessel on their Latvia – Sweden service in the summer. It is a freight service and is operated by the Rederi AB Gotland's *Gute*. The former *Sally Sun* was lengthened by 20.3m in Gdynia, Poland earlier in the year. The new service operates to the coastal port of Nynashamn rather than into Stockholm itself.

Scandlines changed their Rostock – Helsinki – Muuga ro-ro service to run only between Germany and Finland. The three ships used – the Estonian *Lehola* and *Lembitu* and the ex Finnlines *Aurora* – were thus able to offer a six day per week service. In the autumn, Scandlines purchased the *Aurora*, with the intentions that she be renamed and the Polish crew replaced by Germans displaced from the now withdrawn *Rügen*. However, in December it was announced that the Rostock – Helsinki service would cease at the end of the year as they had been unable to secure a satisfactory charter agreement for the Estonian vessels. These two vessels soon found alternative charterers, a reflection of the chronic shortage of modern medium sized ro-ro ships. Mann Lines closed down their Paldiski – Kiel – Århus – Gdansk – Paldiski -service operated by the *Hamburg* in April. The service had been acquired from ESCO the previous year. TransRussiaExpress placed the Stena owned *Vasaland* on their Lübeck – Sassnitz – Baltijsk (Kaliningrad) – St Petersburg service in May. She lasted on the route until January 2005 when she was replaced by the *Baltic Eager*, formerly the *Seawheel Humber*.

NORTH BALTIC

In March, Tallink introduced a 100 driver ro-ro ferry to replace the *Kapella* on the Paldiski – Kapellskär route – the *Regal Star*, previously owned by MCL of Italy and, since the autumn of 2000 operating between Salerno, Palermo and Valencia on charter to Grimaldi Ferries. She has a rather unusual history, having being laid down as a deep-sea ro-ro at the Severnaya Wharf shipyard in St Petersburg in 1992 but never completed. In 1999 she was purchased by the Paulmba shipyard in Naples who finished her in her present form. She offers twice the vehicle capacity of the *Kapella*, which is was transferred to offer additional freight capacity between Helsinki and Tallinn. Her 80 passenger capacity was uprated to 100 by re-designating some crew cabins for drivers as she runs with a smaller crew than in the Mediterranean.

Tallink's second newbuilding arrived in April. The new *Victoria I* (a near sister of the *Romantika*) was placed on the Tallinn – Stockholm route. Following Estonia's joining the EU in May a Mariehamn call was introduced on this route so duty free sales could continue. Also in April, Tallink inaugurated a new triangular route Helsinki – Tallinn – St Petersburg – Helsinki service with the *Fantaasia*. The service ended at the end of the year and the *Fantaasia* went on a nine month charter o Algérie Ferries. During 2004, Tallink purchased two new fast ferries – the *Shannon Alexis* (once the *Stena Pegasus* with Stena Line on Newhaven – Dieppe) and *St Matthew* (once the *Pegasus Two* with Color Line on Larvik – Skagen).

Both were Fincantieri MDV1200 monohull vessels, precursors of Sea Containers' SuperSeaCats but 5m shorter and having hulls of high tensile steel rather than aluminium. They had chequered careers and were among some of the least successful fast-ferries, having been laid up in Gibraltar for several years. They were renamed the *Tallink Autoexpress 3* and *Tallink Autoexpress 4* .

In the autumn, Tallink ordered another cruise ferry from Aker Finnyards for delivery in 2006. Some 18m longer than the *Romantika* and *Victoria I* and carrying up to 2800 passengers, she will be placed on the Tallinn – Helsinki route, enabling the *Romantika* to join her near sister on the Tallinn – Stockholm route.

Eckerö Line took delivery in May of a second vessel for their Helsinki – Tallinn service – the ro-pax *Transparaden*, which had served on both Harwich – Hoek van Holland and Ramsgate – Ostend routes in the 1990s. She was renamed the *Translandia* and provide freight back-up to the passenger vessel *Nordlandia*. Since December 2004 she had been operating between St Petersburg and Kiel for SCF St Petersburg Line and had enabled the company to expand into the passenger market. However, her sale forced the company back into pure freight operation as no suitable alternative tonnage was available.

Since the opening up of Estonia following that country's independence in 1989, services between Finland and Estonia had been very much based on the duty free market and, to some extend, the need to be out of Finland for 22 hours or more to qualify for the full duty free allowance. Following Estonia's accession to the EU, duty paid shopping came to the fore (although Finland's decision to cut their very high alcohol taxes by 30% had some effect on expected traffic levels). Tallink's *Meloodia* switched to a two crossings per day ferry schedule, leaving only the *Romantika* on the traditional cruise schedule. Eckerö Line's *Nordlandia* – operating day trips rather than 22 hour sailings – lost vehicle deck capacity following the building of a large duty paid shop.

In August, Estonian company Saaremaa Laevakompanii and Tallinna Sadam (Port of Tallinn Authority) agreed to operate joint-venture services to the Estonian Baltic Sea islands. Economy Minister Meelis Atonen rejected this however because subsidy would have been higher than previously and issued orders that Saaremaa Laevakompanii should continue to operate services with the old subsidies for 2_ years from October 2004. However, this rather Soviet style direction of a private company did not go down well in democratic Estonia and he was over-ruled in the ensuing controversy. The joint venture then proceeded. In addition, the company announced plans to operate a new route between Saaremaa and Ventspils in Latvia. It is likely to be introduced in May 2005.

After an autumn spent on cruises from Stockholm, Silja Line's passenger only *Silja Opera* switched to operating a daily service between Helsinki and Tallinn. She was due switch back to longer cruises when the fast ferries *SuperSeaCat Three* and *SuperSeaCat Four* started again in the spring.

In the autumn, Finnlines ordered two more 216m 500 passenger vessels from Fincantieri. However, whilst the first three builds were to operate for Finnlines between Travemünde and Helsinki, these vessels, due to be delivered in 2007, will operate on the FinnLink service between Naantali and Kapellskär, allowing the existing vessels to be deployed on the Nordö-Link service from Travemünde to Malmö.

Eckerö Linjen purchased the 1979 built *Jens Kofoed* from BornholmsTrafikken. She is due to be delivered in May when the Danish company take delivery of their new ships *Dueodde* and *Hammerodde*. There is a condition of the sale that, should their delivery be delayed, the *Jens Kofoed* will be retained. She is to undergo a major rebuilt to make her suitable for North Baltic operation and is likely to replace one the two 'Papenburg sisters' owned by the company, the *Alandia* and *Roslagen*, in late 2005 or early 2006

In the autumn, RG-Line purchased the Polferries ship *Kahleberg* from her German owners. Delivery was made at the end of 2004 but she had to be modified to increase passenger capacity from 79 to 300. Meanwhile, RG-Line's existing ship, the *Casino Express*, ran aground off Uumaja on 24th November and was badly damaged. RG-Line then chartered Eckerö Linjen's *Alandia* (laid up for the winter) as a temporary replacement but on 8th December the *Alandia* ran aground on almost the same spot and damaged her propeller. In the interim RG-Lines operated an overnight replacement bus service round the north of the Gulf of Bothnia – a distance of 513 miles. Later they operated the *Kahleberg* (in un-rebuilt form), the *Alandia* and a repaired *Casino Express*.

Nick Widdows

Hammerodde *(Ton Grootenboer)*

Robin Hood *(FERRYInformation)*

Bergen
Stavanger
Egersund
Kristiansand
Hanstholm
Kalundborg
Esbjerg
Cuxhaven
Norddeich
Harlingen
Emden
North Sea
U. K.
IJmuiden
Scheveningen
Hoek van Holland
NETH.
Rotterdam
Harwich
Plymouth
Portsmouth
Ramsgate
Dover
Newhaven
Vlissingen
Ostend
Zeebrugge
Calais
Dunkerque
English Channel
BELGIUM
Cherbourg
Dieppe
Roscoff
Le Havre
Caen
LUX.
St Malo
FRANCE

FINLAND
Rauma
Naantali
Turku
Kotka
Helsinki
ÅLAND
Mariehamn
Långnäs
Hanko
Oslo
SWEDEN
Kapellskär
Tallinn
Paldiski
Moss
Stockholm
Södertälje
Strömstad
Nynäshamn
ESTONIA
Västervik
Visby
Ventspils
Riga
Oskarshamn
Varberg
GOTLAND
LATVIA
ÖLAND
Liepaja
Helsingborg
Karlskrona
Karlshamn
Klaìpeda
Malmö
Ystad
LITHUANIA
Trelleborg
Rønne
BORNHOLM
Gedser
Sassnitz
Gdynia
Rostock
Gdansk
BYELA
Swinoujscie
POLAND
UKR
CZECH
SLOVAKIA

PORTSMOUTH QUEEN

Portsmouth Queen & Val de Loire *(Miles Cowsill)*

section I

gb & ireland

BRITTANY FERRIES

THE COMPANY *Brittany Ferries* is the trading name of *BAI SA*, a French private sector company and the operating arm of the *Brittany Ferries Group*. The UK operations are run by *BAI (UK) Ltd*, a UK private sector company, wholly owned by the *Brittany Ferries Group*.

MANAGEMENT Group Managing Director Michel Maraval, **Managing Director UK & Ireland** David Longden.

ADDRESS Millbay Docks, Plymouth, Devon PL1 3EW.

TELEPHONE Administration ***Plymouth*** +44 (0)8709 010500, ***Portsmouth*** +44 (0)8709 011300, **Reservations** ***All Services*** +44 (0)8705 360360.

FAX Administration & Reservations +44 (0)8709 011100.

INTERNET Website www.brittanyferries.com *(English, French, Spanish, German)*

ROUTES OPERATED Conventional Ferries ***All year*** Plymouth – Roscoff (6 hrs (day), 6 hrs – 7 hrs 30 mins (night); *(2,7)*; up to 3 per day (summer), 1 per day (winter)), Portsmouth – St Malo (8 hrs 45 mins (day), 10 hrs 30 mins – 11 hrs 30 mins (night); *(8))*; 1per day), Portsmouth – Caen (Ouistreham) (6 hrs (day), 6 hrs 15 mins – 8 hrs (night); *(4,5)*; 3 per day), Plymouth – Santander (Spain) (18 hrs; *(7)*; 2 per week (March – November)), Poole – Cherbourg (4 hrs 15 mins; *(1)*; up to 2 per day), Portsmouth – Cherbourg (4 hrs 45 mins; (*1*); 1 per day), ***Summer only*** Cork – Roscoff (14 hrs; *(7)*; 1 per week). **Fast Ferries All year** Portsmouth – Cherbourg (3 hrs; (*6*); 1 or 2 per day), **Summer only** Poole – Cherbourg (2 hrs 15 mins; *(CONDOR VITESSE of Condor Ferries)*; 1 per day), Portsmouth – Caen (Ouistreham) (3 hrs 45 mins (*6*); 3 per week). **Note** The Poole – Cherbourg fast ferry service is operated from May to September by *Condor Ferries* jointly with *Brittany Ferries*.

1	BARFLEUR	20133t	92	19.3k	158.0m	1212P	590C	112T	BA2	FR
2	BRETAGNE	24534t	89	21.0k	151.0m	1926P	580C	84T	BA	FR
4	MONT ST MICHEL	35592t	02	21.5k	173.0m	2200P	880C	166T	BA2	FR
5	NORMANDIE	27541t	92	20.5k	161.0m	2120P	600C	126T	BA2	FR
6»	NORMANDIE EXPRESS	6581t	00	42.0k	97.2m	900P	260C	-	A	BS
7	PONT-AVEN	41700t	04	27.0k	184.3m	2400P	650C	85L	BA	FR
8	VAL DE LOIRE	31788t	87	20.0k	162.0m	2140P	600C	104T	BA	FR

BARFLEUR Built by Kvaerner Masa-Yards, Helsinki for the *Truckline* Poole – Cherbourg service to replace two passenger vessels and to inaugurate a year round passenger service. In 1999 the *Truckline* branding was dropped for passenger services and she was repainted into full *Brittany Ferries* livery.

BRETAGNE Built by Chantiers De l'Atlantique, St Nazaire for the Plymouth – Santander and Cork – Roscoff services (with two sailings per week between Plymouth and Roscoff). In 1993 she was transferred to the Portsmouth – St Malo service. In 2004 also operated between Portsmouth and Cherbourg. In 2005 to operate between Plymouth and Roscoff.

MONT ST MICHEL Built by Van der Giessen-de Noord, Krimpen aan den IJssel, Rotterdam for *Brittany Ferries* to replace the DUC DE NORMANDIE on the Portsmouth – Caen route.

Brittany Ferries
NEW HIGH SPEED SERVICES
More choice to Holiday France & Spain
In addition to our existing cruise and high speed ferry services, we now offer **2 new high speed routes** from Portsmouth to Cherbourg and Caen.
As part of Brittany Ferries on-going investment into our fleet and services we aim to continue to provide the best choice and value for cross-channel travellers.
Visit **brittanyferries.com** or call **0870 908 9682**
Brittany Ferries
CORK
PLYMOUTH
POOLE
PORTSMOU
ROSCOFF
ST MALO
CHERBOURG
CAEN
SANTANDER

Pont-Aven *(Miles Cowsill)*

Condor Express *(Brian Mitchell)*

NORMANDIE Built by Kvaerner Masa-Yards, Turku, Finland for the Portsmouth – Caen route.

NORMANDIE EXPRESS InCat Evolution 10 catamaran built as the INCAT TASMANIA. In November 2000 chartered to *TranzRail* of New Zealand and renamed THE LYNX. Placed on the Wellington – Picton service. In July 2003 replaced by 1997 built InCat 86m craft INCAT 046 given the marketing name 'The Lynx' and laid up. In spring 2005 chartered to *Brittany Ferries* to operate on their Cherbourg – Portsmouth and Caen – Portsmouth services and renamed the NORMANDIE EXPRESS.

PONT-AVEN Built by Jos L Meyer Werft, Papenburg, Germany to operate on the Plymouth – Roscoff, Plymouth – Santander and Cork – Roscoff routes.

VAL DE LOIRE Built by Schichau Seebeckwerft AG, Bremerhaven, Germany as the NILS HOLGERSSON for *Rederi AB Swedcarrier* of Sweden for their service between Trelleborg and Travemünde, joint with *TT-Line* of Germany (trading as *TT-Line*). In 1992 purchased by *Brittany Ferries* for entry into service in spring 1993. After a major rebuild, she was renamed the VAL DE LOIRE and introduced onto the Plymouth – Roscoff, Plymouth – Santander and Cork – Roscoff routes. In 2004 transferred to the Portsmouth – St Malo and Portsmouth – Cherbourg services. In 2005 to operate mainly Portsmouth – St Malo.

CELTIC LINK FERRIES

THE COMPANY *Celtic Link Ferries Ltd* is an Irish Republic private sector company.

MANAGEMENT Marketing Manager Paul Tyrrell.

ADDRESS Kilmore Quay, Co Wexford, Republic of Ireland.

TELEPHONE Administration +353 (0)1 855 8268, **Reservations *Passenger* UK** +44 (0)870 458 0401, **Irish Republic** +353 (0)1 855 8268, ***Freight*** +353 (0)1 855 8246 or 855 8268.

FAX Administration & Reservations +353 (0)1 855 8203.

INTERNET Website www.celticlinkferries.com *(English)*

ROUTES OPERATED Rosslare – Cherbourg (18 hrs; *(1)*; 3 per week).

1	DIPLOMAT	16776t	78	17.0k	151.0m	74P	-	122T	A2	BM

DIPLOMAT Built by Hyundai Shipbuilders & Heavy Industries, Ulsan, South Korea as the STENA TRANSPORTER, for *Stena Rederi* of Sweden. In 1979 she was renamed the FINNROSE and chartered to *Finnlines*. She later served with *Atlanticargo* on their service between Europe and USA/Mexico. In 1980 she returned to *Stena Line* and resumed her original name. Later in 1980 she was chartered to *European Ferries* for their Felixstowe – Rotterdam freight-only service and renamed the BALTIC FERRY. In 1982 she served in the Falkland Islands Task Force. In 1986 she was converted to ro-pax format and moved to the Felixstowe – Zeebrugge passenger service. In 1992 she was renamed the PRIDE OF SUFFOLK. In 1994 she was purchased by *P&O European Ferries*. In 1995 the Felixstowe – Zeebrugge passenger service ceased, most of her additional passenger accommodation was removed, passenger capacity was reduced and she was transferred to the Felixstowe – Rotterdam freight service. In 2001 transferred to the *P&O Irish Sea's* Liverpool – Dublin route and renamed the EUROPEAN DIPLOMAT. In 2002 transferred to the Rosslare – Cherbourg route; this service ended in December 2004. In 2004 purchased by *Celtic Link Ferries*, renamed the DIPLOMAT and in February 2005 re-launched the Rosslare – Cherbourg link.

CONDOR FERRIES

THE COMPANY *Condor Ferries Ltd* is a Channel Islands private sector company owned by the *Condor Group*, Guernsey which is owned by the *Royal Bank of Scotland*.

MANAGEMENT Managing Director Robert Provan, **Freight Director** Jeff Vidamour, **General Manager, Sales & Marketing** Nicholas Dobbs.

ADDRESS Head Office PO Box 10, New Jetty Offices, White Rock, St Peter Port, Guernsey GY1 3AF, Sales and Marketing Condor House, New Harbour Road South, Hamworthy, POOLE BH15 4AJ.

TELEPHONE Administration *Guernsey* +44 (0)1481 728620, ***Poole*** +44 (0)1202 207207, **Reservations** +44 (0)845 345 2000.

FAX Administration *Guernsey* +44 (0)1481 728521, ***Poole*** +44 (0)1202 685184, **Reservations** +44(0)1305 760776.

INTERNET Email reservations@condorferries.co.uk **Website** www.condorferries.com *(English)*

ROUTES OPERATED Conventional Ferry *All year* Portsmouth – St Peter Port (Guernsey) (6 hrs 30 mins) – St Helier (Jersey) (10 hrs 30 mins) – Portsmouth (8 hrs 30 mins) (return Guernsey via Jersey, 12 hrs 30 mins); *(1)*; 6 per week, ***Summer only*** Portsmouth – Cherbourg (France) (5 hrs; *(1)*; 1 per week). **Fast Ferries *Winter Only*** Weymouth – St Peter Port (Guernsey) (2 hrs) – St Helier (Jersey via Guernsey) (3 hrs 20 mins); *(3)*; twice weekly), ***Spring and Autumn*** Weymouth – St Peter Port (Guernsey) (2 hrs) – St Helier (Jersey via Guernsey) (3 hrs 20 mins); *(3)*; 1 per day, Poole – St Peter Port (Guernsey) (2 hrs 30 mins) – St Helier (Jersey via Guernsey) (3 hrs 50 mins); *(3)*; daily, ***Summer*** Weymouth – St Peter Port (Guernsey) (2 hrs) – St Helier (Jersey via Guernsey) (3 hrs 20 mins); *(3)*; 1 per day, Poole – St Peter Port (Guernsey) (2 hrs 30 mins) – St Helier (via Guernsey 3 hrs 50 mins, direct 3 hrs) – St Malo (4 hrs 35 min) ; *(3,4)*; up to 2 per day to Jersey and Guernsey, 1 per day to St Malo (***Note*** Poole – St Malo service operates via either Guernsey or Jersey), St Malo – St Peter Port (Guernsey) (2hrs 55 mins via Jersey, 1hr 45 mins direct; *(2)*; 2 per day), St Malo – St Helier (Jersey) (1 hr 10 mins; *(2)*; 1 per day – no cars conveyed on this service), Poole – Cherbourg (2 hrs 15 mins; *(4)*; 1 per day (service operated jointly with *Brittany Ferries)*).

1	COMMODORE CLIPPER	14000t	99	18.25k	129.1m	500P	100C	92T	A	BS
2»	CONDOR 10	3241t	93	37.0k	74.3m	580P	80C	-	BA	BS
3»	CONDOR EXPRESS	5005t	96	39.0k	86.6m	774P	185C	-	A2	BS
4»	CONDOR VITESSE	5005t	97	39.0k	86.6m	774P	185C	-	A2	BS

COMMODORE CLIPPER Ro-pax vessel built by Van der Giessen-de Noord, Krimpen aan den IJssel, Rotterdam for *Commodore Ferries* to operate between Portsmouth and the Channel Islands. She replaced the ISLAND COMMODORE, a freight only vessel. Her passenger capacity is normally restricted to 300 but is increased to 500 when the fast ferries are unable to operate.

CONDOR 10 InCat 74m catamaran. Built by Hobart, Australia for the *Holyman Group* for use by *Condor Ferries*. In summer 1995 she was chartered to *Viking Line* to operate between Helsinki and Tallinn under the name 'VIKING EXPRESS II' (although not officially renamed). In summer 1996 she was chartered to *Stena Line* to operate between Fishguard and Rosslare. During northern hemisphere winters she served for *TranzRail* of New Zealand for the service between Wellington (North Island) and Picton (South Island). In May 1997 she was transferred to *Holyman Sally Ferries* and inaugurated a new Ramsgate – Dunkerque (Est) service, replacing the Ramsgate – Dunkerque (Ouest) service of *Sally Ferries*. After further service in New Zealand during winter 1997/98, in summer 1998 she was due to operate between Weymouth, Guernsey and St Malo, but, in the event, the CONDOR VITESSE was chartered for that route and she was laid up in Australia. Refurbished during 2001, she returned to the UK in spring 2002 replacing the passenger-only CONDOR 9 between St Malo and Guernsey and Jersey. She is now owned by *Condor Ferries*.

CONDOR EXPRESS InCat 86m catamaran built at Hobart, Tasmania, Australia. She was delivered December 1996 and entered service in 1997.

CONDOR VITESSE InCat 86m catamaran built at Hobart. Built speculatively and launched as the INCAT 044. Moved to Europe in summer 1997 and spent time in the both the UK and Denmark but was not used. In 1998, she was chartered to *Condor Ferries* and renamed the CONDOR VITESSE. During winter 1999/2000 she was chartered to *TranzRail* of New Zealand. Returned to UK in spring 2000.

DFDS SEAWAYS

THE COMPANY *DFDS Seaways* is the passenger division of the *DFDS A/S*, a Danish private sector company. *DFDS Seaways Ltd* is a UK subsidiary.

MANAGEMENT Managing Director DFDS A/S Ole Frie, **Director DFDS Seaways** Thor Johannesen, **Managing Director DFDS Seaways Ltd** John Crummie.

ADDRESS Scandinavia House, Parkeston, Harwich, Essex CO12 4QG.

TELEPHONE Administration +44 (0)1255 243456, **Reservations *National*** 08705 333000 (from UK only), ***Harwich*** +44 (0)1255 240240, ***Newcastle*** +44 (0)191-293 6283.

FAX Administration & Reservations *Harwich* +44 (0)1255 244370, ***Newcastle*** +44 (0)191-293 6245.

INTERNET Email john.crummie@dfds.co.uk **Website** www.dfdsseaways.co.uk *(English)*

www.dfdsseaways.com *(Danish, Dutch, German, Norwegian, Swedish)*

ROUTES OPERATED *All year* Harwich – Esbjerg (Denmark) (17 hrs; *(1)*; 3 per week), Kristiansand (Norway) – Gothenburg (Sweden) (7 hrs (day), 13 hrs 30 min (night); *(4)*; 1 per week (spring, summer and autumn), 3 per week (winter)), Harwich – Cuxhaven (Germany) (16 hrs; *(2)*; 3 per week (winter, spring, autumn), alternate days (summer), Newcastle (North Shields) – IJmuiden (near Amsterdam, Netherlands) (15 hrs; *(3,5)*; daily), ***Spring, Summer and Autumn only*** Newcastle – Kristiansand – Gothenburg (25 hrs; *(4)*; 2 per week).

1	DANA SIRENA	22382t	03	23.0k	199.4m	602P	316C	154T	A	DK
2	DUCHESS OF SCANDINAVIA	16794t	93	19.0k	134.4m	882P	350C	44T	BA	DK
3	DUKE OF SCANDINAVIA	19589t	78	20.5k	152.9m	1120P	375C	45T	BA	DK
4	PRINCESS OF SCANDINAVIA	22528t	76	26.0k	184.6m	1620P	375C	36T	AS	DK
5	QUEEN OF SCANDINAVIA	34093t	81	20.0k	168.1m	1638P	360C	54T	BA	DK

DANA SIRENA Built by Stocznia Szczecinska, Szczecin, Poland for *Lloyd Sardegna* of Italy as the GOLFO DEI DELFINI for service between Italy and Sardinia. However, due to late delivery the order was cancelled. In 2002 purchased by *DFDS Seaways*, and, during winter 2002/3, passenger accommodation was enlarged and refitted, increasing passenger capacity from 308 to 600. In June 2003, renamed the DANA SIRENA, she replaced unmodified sister vessel, the DANA GLORIA (now with *Lisco Baltic Services* – see Section 6), on the Esbjerg – Harwich service. She is branded *DFDS Tor Line*.

DUCHESS OF SCANDINAVIA Built by Fosen Mekaniske Verksted, Rissa, Norway as the BERGEN for *Rutelaget Askøy-Bergen* and used on the *Fjord Line* Bergen (Norway) – Egersund (Norway) – Hanstholm (Denmark) service. In 2003 chartered to *DFDS* and in April 2003, renamed the DUCHESS OF SCANDINAVIA and, after modifications, introduced onto the Harwich – Cuxhaven service. In 2004 sold to *Bergensfjord KS* of Norway but continued to be chartered to *DFDS*.

DUKE OF SCANDINAVIA Built by Aalborg Værft A/S, Aalborg, Denmark as the DANA ANGLIA for the Harwich – Esbjerg service and seldom operated elsewhere. In autumn 2002, renamed the DUKE OF SCANDINAVIA and inaugurated a new Copenhagen – Trelleborg – Gdansk service. In 2003 moved to the Newcastle – IJmuiden service.

PRINCESS OF SCANDINAVIA Built by Lübecker Flender-Werke, Lübeck, Germany as the TOR SCANDINAVIA for *Tor Line* of Sweden for their Amsterdam – Gothenburg and Felixstowe – Gothenburg services. In 1979 she was used on a world trade cruise and was temporarily renamed the HOLLAND EXPO. Similar exercises were undertaken in 1980, 1982 and 1984, but on these occasions her temporary name was the WORLD WIDE EXPO. She was acquired by *DFDS* in 1981 and subsequently re-registered in Denmark. She has also operated on the Harwich – Esbjerg service. Between 1989 and 1993 she also operated Newcastle – Esbjerg and Amsterdam – Gothenburg services. In 1991, following a major refurbishment, she was renamed the PRINCESS OF SCANDINAVIA. Since 1994, she generally operated on the Harwich – Gothenburg and Newcastle – Gothenburg routes. During winter 1998/1999 she had major modifications made at Gdansk and

during winter 1999/2000 she had a major engine rebuild. She now operates between Newcastle and Gothenburg via Kristiansand.

QUEEN OF SCANDINAVIA Built by Oy Wärtsilä Ab, Turku, Finland as the FINLANDIA for *EFFOA* of Finland for *Silja Line* services between Helsinki and Stockholm. In 1990 she was sold to *DFDS*, renamed the QUEEN OF SCANDINAVIA and introduced onto the Copenhagen – Helsingborg – Oslo service. In 2001 transferred to the Newcastle – IJmuiden route.

EMERAUDE JERSEY FERRIES

THE COMPANY *Emeraude Jersey Ferries* is a French private sector company, owned by *Sogestran* of France. It took over the operations *Emeraude Lines* in 2004.

MANAGEMENT Channel Islands Manager Gordon Forrest, **Brittany and Normandy Office Manager** Vacant.

ADDRESS Terminal Ferry du Naye, PO Box 16, 35401, St Malo Cedex, France.

TELEPHONE Administration & Reservations *St Malo* +33 (0)2 23 180 180, ***Jersey*** +44 (0)1534 766566.

FAX Administration & Reservations *St Malo* +33 (0) 2 23 181 500, ***Jersey*** +44 (0)1534 768741.

INTERNET Email sales@emeraude.co.uk **Website** www.emeraudeferries.com *(English)*

ROUTES OPERATED Fast Ferry St Malo (France) – St Helier (Jersey) (1 hr 15 mins; *(1)*; up to 3 per day).

1»	SOLIDOR 5	2300t	01	38.0k	60.0m	450P	60C	-	A	FR

SOLIDOR 5 Fjellstrand JumboCat 60m catamaran, built at Omastrand, Norway for *Emeraude Lines* to replace the SOLIDOR 3. Entered service in January 2001. Due to a dispute between the owners of the vessel and *Emeraude Jersey Ferries*, she was laid up during 2004. In March 2005 chartered to *Emeraude Jersey Ferries* whilst the chartered EMERAUDE FRANCE (see *Sea Containers Ferries*) was being refitted. In April returned to full service following resolution of the dispute and the ending of the one year charter agreement with *Sea Containers*.

FJORD LINE

THE COMPANY *North Fjord Line* is a Norwegian company, 100% owned by *Bergen-Nordhordland Rutelag AS (BNR)*.

MANAGEMENT Managing Director (UK) Dag Romslo, **Sales Director (UK)** Mike Wood,

ADDRESS *UK* Royal Quays, North Shields NE29 6EG.

TELEPHONE *UK* Administration +44 (0)191-296 1313, **Reservations** +44 (0)191-296 1313.

FAX +44 (0)191-296 1540, **Telex** 537275.

INTERNET Email fjordline.uk@fjordline.com **Website** www.fjordline.co.uk *(English)*

ROUTES OPERATED Bergen (Norway) – Haugesund (Norway) – Stavanger (Norway) – Newcastle – Bergen (triangular route), Bergen – Haugesund – Stavanger – Newcastle (Bergen – Stavanger (via 6 hrs), Stavanger – Newcastle (direct 18 hrs 30 mins, via Bergen 29 hrs 30 mins), Bergen – Newcastle (21 hrs 15 mins); *(1)*; 3 sailings Norway – Newcastle per week).

Fjord Line also operates between Norway and Denmark; see Section 6.

1	JUPITER	20581t	75	21.0k	175.3m	1250P	285C	40T	BA	NO

JUPITER Built by Dubegion-Normandie SA Nantes, France as the WELLAMO for *EFFOA* of Finland for *Silja Line* services between Helsinki and Stockholm. In 1981 sold to *DFDS*, renamed the DANA GLORIA and placed onto the Gothenburg – Newcastle and Esbjerg – Newcastle services. In 1983 she was moved to the Copenhagen – Oslo service. In 1984 she was chartered to *Johnson Line* of Sweden for *Silja Line* service between Stockholm and Turku and renamed the SVEA CORONA. This charter

Duke of Scandinavia *(Miles Cowsill)*

Isle of Innishmore *(FotoFlite)*

ended in 1985 and she returned to the Copenhagen – Oslo service and resumed the name DANA GLORIA. During winter 1988/89 she was lengthened by Meyer Werft, Papenburg, Germany and in early 1989 she was renamed the KING OF SCANDINAVIA. She returned to the Copenhagen – Oslo route; in 1990 a Helsingborg call was introduced. In 1994 she was sold to *Color Line* and renamed the COLOR VIKING. In 1998 she was sold to *Fjord Line* and renamed the JUPITER.

IRISH FERRIES

THE COMPANY *Irish Ferries* is an Irish Republic private sector company, part of the *Irish Continental Group*. It was originally mainly owned by the state owned *Irish Shipping* and partly by *Lion Ferry AB* of Sweden. *Lion Ferry* participation ceased in 1977 and the company was sold into the private sector in 1987. Formerly state owned *B&I Line* was taken over in 1991 and from 1995 all operations were marketed as *Irish Ferries*.

MANAGEMENT Group Managing Director Eamon Rothwell, **Group Marketing Director** Tony Kelly.

ADDRESS 2 Merrion Row, Dublin 2, Republic of Ireland.

TELEPHONE Administration +353 (0)1 855 2222, **Reservations *Dublin*** +353 (0)1 638 3333, ***Cork*** +353 (0)21 455 1995, ***Rosslare Harbour*** +353 (0)53 33158, ***Holyhead*** 0990 329129 (from UK only), ***Pembroke Dock*** 0990 329543 (from UK only), ***National*** 08705 171717 (from UK only), ***24 hour information*** +353 (0)1 661 0715.

FAX Administration & Reservations *Dublin* +353 (0)1 661 0743, ***Cork*** +353 (0)21 450 4651, ***Rosslare*** +353 (0)53 33544.

INTERNET Email *Britain* info@irishferries.co.uk ***Elsewhere*** info@irishferries.com **Website** www.irishferries.com *(English)*

ROUTES OPERATED Conventional Ferries Dublin – Holyhead (3 hrs 15 mins; *(4)*; 2 per day), Rosslare – Pembroke Dock (3 hrs 45 mins; *(1)*; 2 per day), Rosslare – Cherbourg (France) (17 hrs 30 mins; *(3)*; 1 or 2 per week), Rosslare – Roscoff (France) (16 hrs; *(3)*; 1 or 2 per week). **Note** the Rosslare – Cherbourg/Roscoff service operates on a seasonal basis. **Fast Ferry** Dublin – Holyhead (1 hr 49 min; *(2)*; up to 4 per day). Marketed as 'DUBLIN*Swift*'.

1	ISLE OF INISHMORE	34031t	97	21.5k	182.5m	2200P	802C	152T	BA2	IR
2»	JONATHAN SWIFT	5989t	99	39.5k	86.6m	800P	200C	-	BA	IR
3	NORMANDY	24872t	82	19.0k	149.0m	1526P	420C	62L	BA2	BS
4	ULYSSES	50938t	01	22.0k	209.0m	1875P	1342C	300T	BA2	IR

ISLE OF INISHMORE Built by Van der Giessen-de Noord, Krimpen aan den IJssel, Rotterdam for *Irish Ferries* to operate on the Holyhead – Dublin service. In 2001 replaced by the ULYSSES and moved to the Rosslare – Pembroke Dock route. She also relives on the Dublin – Holyhead route when the ULYSSES receives her annual overhaul.

JONATHAN SWIFT Austal Auto-Express 86 catamaran built at Fremantle, Australia for *Irish Ferries* for the Dublin – Holyhead route.

NORMANDY Built by Götaverken Arendal AB, Gothenburg, Sweden. One of two vessels ordered by *Göteborg-Frederikshavn-Linjen* of Sweden (trading as *Sessan Linjen*) before the take over of their operations by *Stena Line AB* in 1981. Both were designed for the Gothenburg – Frederikshavn route (a journey of about three hours). However, *Stena Line* decided in 1982 to switch the first vessel, the KRONPRINSESSAN VICTORIA (now the STENA EUROPE of *Stena Line AB*), to their Gothenburg – Kiel (Germany) route since their own new tonnage for this route, being built in Poland, had been substantially delayed. She was modified to make her more suitable for this overnight route. Work on the second vessel – provisionally called the DROTTNING SILVIA – was suspended for a time but she was eventually delivered, as designed, in late 1982 and introduced onto the Gothenburg – Frederikshavn route on a temporary basis pending delivery of new *Stena Line* ordered vessels. She was named the PRINSESSAN BIRGITTA, the existing ex *Sessan Linjen* vessel of the same name being renamed the STENA SCANDINAVICA. In early 1983 she was substantially modified in a similar way to her sister. In June 1983 she was renamed the ST NICHOLAS, re-registered in Great Britain and

IRISH FERRIES
Pay Economy
Travel First Class
IRISH FERRIES
DUBLIN Swift
IRISH FERRIES
DUBLIN
HOLYHEAD
ROSSLARE
PEMBROKE
CHERBOURG
ROSCOFF
With the most luxurious fleet on the Irish Sea, Irish Ferries is the very first company offering low fare travel you can actually enjoy! We also offer the widest range of great value package holiday deals to Ireland, Britain & France. Book on-line at our award winning website www.irishferries.com for our very lowest fares.
IrishFerries.com
You'll get our best deals on the web!
UK 08705 17 17 17 ROI 0818 300 400 NI 00353 818 300 400
K185

Dawn Merchant *(Miles Cowsill)*

Lady of Mann *(Miles Cowsill)*

entered service on five year charter to *Sealink UK* on the Harwich – Hoek van Holland route. In 1988 she was purchased and re-registered in The Bahamas. In 1989 she was sold to *Rederi AB Gotland* of Sweden and then chartered back. In 1991 she was renamed the STENA NORMANDY and inaugurated a new service between Southampton and Cherbourg. She was withdrawn in December 1996, returned to *Rederi AB Gotland* and renamed the NORMANDY. In 1997 she was chartered to *Tallink* and operated between Helsinki and Tallinn; this charter ended at the end of the year. In 1998 she was chartered to *Irish Ferries*. She briefly operated between Rosslare and Pembroke Dock before switching to the their French services. In 1999 she was purchased by *Irish Ferries*. She also operates between Rosslare and Pembroke Dock when the regular vessel is away on overhaul. In 2005 transferred to Bahamas flag.

ULYSSES Built by Aker Finnyards, Rauma, Finland for *Irish Ferries* for the Dublin – Holyhead service.

IRISH SEA EXPRESS

THE COMPANY *Irish Sea Express Ltd* is a UK private sector company.

MANAGEMENT Chief Executive Andy Boardman.

ADDRESS Gibraltar House, Kelvin Road, Wallasey, Wirral CH44 7JW.

TELEPHONE Reservations 0870 11 22 222.

FAX Administration +44 (0)151 630 8451.

INTERNET Email reservations@irishseaexpress.com **Website** www.irishseaexpress.com *(English)*

ROUTE OPERATED ***March – December*** Liverpool – Dublin (4 hrs; (*1*); 1/2 per day). Service due to start in April 2005.

1»	SEA EXPRESS 1	3003t	91	37.0k	74.3m	516P	80C	-	BA	UK

SEA EXPRESS 1 InCat 74m catamaran built at Hobart, Tasmania, Australia. Built as the HOVERSPEED FRANCE, the second SeaCat. She inaugurated Dover – Calais/Boulogne service in 1991. In 1992 she was chartered to *Sardinia Express* of Italy and renamed the SARDEGNA EXPRESS; she did not operate on the Channel that year. This charter was terminated at the end of 1992 and in 1993 she was renamed the SEACAT BOULOGNE and operated on the Dover – Calais and Folkestone – Boulogne services. In 1994 she was chartered to *IOMSP Co*, renamed the SEACAT ISLE OF MAN and replaced the LADY OF MANN on services between Douglas (Isle of Man) and Britain and Ireland. During winter 1994/5 operated for *SeaCat Scotland* between Stranraer and Belfast. She returned to *IOMSP Co* in June 1995. During spring 1995 she was chartered to *Condor Ferries*; she then was chartered again to *IOMSP Co* and returned to *Sea Containers* in the autumn. In 1996 she was chartered to *ColorSeaCat KS*, renamed the SEACAT NORGE and inaugurated a new service between Langesund (Norway) and Frederikshavn (Denmark). During winter 1996/97 she operated between Dover and Calais. In early 1997 she was again renamed the SEACAT ISLE OF MAN. During summer 1997 she operated for *IOMSP Co*, serving on Liverpool, Dublin and Belfast seasonal services to Douglas (May to September) plus a weekly Liverpool – Dublin service when the LADY OF MANN operated from Fleetwood. In late 1997 she was transferred to the *Hoverspeed* Dover – Calais route and operated on this route throughout 1998. In 1999 she operated between Douglas and Liverpool and Douglas and Dublin for *IOMSP Co*. In 2000, 2001 and 2002 she also operated between Douglas and Belfast and Douglas and Heysham (plus provide some additional sailings between Liverpool and Dublin in 2002). In early 2003 she replaced the SEACAT SCOTLAND on the Belfast – Troon service. Returned to Isle of Man services in March 2003. In 2003 transferred to the *IOMSP Co*. At end of 2004 summer season laid up. In 2005 chartered to *Irish Sea Express* and renamed the SEA EXPRESS 1.

STEAM PACKET COMPANY

ISLE OF MAN STEAM PACKET COMPANY

THE COMPANY The *Isle of Man Steam Packet Company Limited*, trading as *Steam Packet Company*, is an Isle of Man registered company owned by *Montagu Private Equity Limited* of the UK. It was purchased from *Sea Containers Ferries Ltd* in 2003.

MANAGEMENT Managing Director Hamish Ross, **Operations Director** Mark Woodward.

ADDRESS Imperial Buildings, Douglas, Isle of Man IM1 2BY.

TELEPHONE Administration +44 (0)1624 645645, **Reservations *From UK*** 08705 523523, ***From elsewhere*** +44 (0)1624 661661.

FAX Administration +44 (0)1624 645609.

INTERNET Email spc@steam-packet.com **Website** www.steam-packet.com *(English)*

ROUTES OPERATED Conventional Ferries *All year* Douglas (Isle of Man) – Heysham (3 hrs 30 mins; *(1,2 (during TT races only))*; up to 2 per day), ***Winter only*** Douglas – Liverpool (4 hrs; *(2)*; 1 per day), **Fast Ferries** Douglas – Liverpool (2 hrs 30 mins; *(3)*; up to 3 per day), Douglas – Belfast (2 hrs 45 mins; *(3)*; up to 3 per week), Douglas – Dublin (2 hrs 45 mins; *(3)*; up to 3 per week), Douglas – Heysham (2 hrs; *(3)*; occasional), Douglas – Larne (2hrs 45 mins; *(EXPRESS of P&O Irish Sea)*; extra sailings during TT period only).

1	BEN-MY-CHREE	12504t	98	19.0k	124.9m	630P	-	90T	A	IM
2	LADY OF MANN	4482t	76	21.0k	104.5m	800P	120C	0T	S	IM
3»	SUPERSEACAT TWO	4462t	97	38.0k	100.0m	694P	140C	-	A	UK

BEN-MY-CHREE Built by Van der Giessen-de Noord, Krimpen aan den IJssel, Rotterdam for the *IOMSP Co* and operates between Douglas and Heysham. Additional passenger accommodation was added at her Spring 2004 refit. In 2005 passenger certificate was increased from 500 to 630.

LADY OF MANN Built by Ailsa Shipbuilding Co Ltd, Troon, UK for the *IOMSP Co*. Cars and small vans are side loaded but no ro-ro freight is conveyed. In 1994 replaced by the SEACAT ISLE OF MAN and laid up for sale. She was used in 1995 during the period of the 'TT' motor cycle races between 26th May to 12th June. Later in 1995 she was chartered to *Porto Santo Line* of Madeira for a service from Funchal to Porto Santo. In 1996 she operated throughout the summer, as no SeaCat was chartered. In 1997, she operated for the TT races and then inaugurated a new Liverpool – Dublin service in June, with a weekly Fleetwood – Douglas service until replaced by the SUPERSEACAT TWO in March 1998. In summer 1998, 2000, 2002 and 2003 she operated during the TT race period, plus a number of special cruises and was then chartered to *Acor Line* for service in the Azores. In 2001 the TT was cancelled, as were her extra sailings and cruises, but the charter went ahead. During winter 1998/99 she provided back-up to the fast ferries on the Liverpool – Dublin and Douglas – Liverpool routes. Between November 2000 and February 2001 she operated on the Liverpool – Douglas route and this was repeated between and the same periods 2001/2, 2002/3 and 2003/4. In summer 2002 she was again chartered to *Acor Line*. This was repeated in 2003 and 2004 and is likely to happen again in 2005.

SUPERSEACAT TWO Fincantieri MDV1200 monohull vessel built at Riva Trigoso, Italy. Built for *Sea Containers*. In 1997 operated on the *Hoverspeed* Dover – Calais route. She was withdrawn from this route at the end of 1997 and in March 1998, she inaugurated a Liverpool – Dublin fast ferry service, operated by *IOMSP Co*. In summer 1999 she operated for *Hoverspeed* between Newhaven and Dieppe. In 2000 she returned to the Irish Sea, operating on the Belfast – Heysham service. In summer 2001 she operated between Dover and Calais and Dover and Ostend. In 2002 she was laid up for sale or charter. Since 2003 she has been on charter to the *IOMSP Co*, operating on the Liverpool and Douglas and Liverpool and Dublin services. In 2005 to operate on all fast ferry services, except Liverpool – Dublin service which will not be operated by the company.

Rotterdam service of *North Sea Ferries* (jointly owned by *P&O* and *The Royal Nedlloyd Group* of The Netherlands until 1996). In December 2001, she was replaced by the new PRIDE OF HULL and, after a two month refurbishment, in 2002 transferred to the Hull – Zeebrugge service, replacing the NORLAND (26290t, 1974). In 2003 renamed the PRIDE OF YORK.

P&O IRISH SEA

THE COMPANY *P&O Irish Sea* is the trading name of *P&O European Ferries (Irish Sea) Ltd*, a British private sector company and a subsidiary of the *Peninsular and Oriental Steam Navigation Company*. It was formed in 1998 by the merger of the shipping activities *Pandoro Ltd* and the Cairnryan – Larne services of *P&O European Ferries (Felixstowe) Ltd.*

MANAGEMENT Chairman Russ Peters, **Managing Director** J H Kearsley, **Passenger Services – Sales and Marketing Manager** James Esler.

ADDRESS Compass House, Dock Street, Fleetwood, Lancashire FY7 6HP.

TELEPHONE Administration +44 (0)1253 615700, **Reservations *UK*** 08702 424777, ***Irish Republic*** 1 800 409 049.

FAX Administration & Reservations +44 (0)1253 615740.

INTERNET Website www.poirishsea.com *(English)*

ROUTES OPERATED Conventional Ferries Cairnryan – Larne (1 hr 45 min; *(1,2)*; 7 per day), Liverpool – Dublin (8 hrs; *(4,5)*; up to 2 per day). **Fast Ferry** (April – September) Cairnryan – Larne (1 hr; *(3)*; 1 per day), Troon – Larne (1 hr 49 min; *(3)*; 2 per day).

1	EUROPEAN CAUSEWAY	20800t	00	23.0k	159.5m	410P	375C	107T	BA2	BS
2	EUROPEAN HIGHLANDER	20800t	02	23.0k	159.5m	410P	375C	107T	BA2	BS
3»	EXPRESS	5902t	98	41.0k	91.3m	920P	225C	-	A	BS
4	NORBANK	17464t	93	22.0k	166.7m	114P	-	150T	A	NL
5	NORBAY	17464t	94	22.0k	166.7m	114P	-	150T	A	UK

EUROPEAN CAUSEWAY Built by Mitsubishi Heavy Industries, Shimonoeki, Japan for *P&O Irish Sea* for the Cairnryan – Larne service.

EUROPEAN HIGHLANDER Built by Mitsubishi Heavy Industries, Shimonoeki, Japan for *P&O Irish Sea* for the Cairnryan – Larne service.

EXPRESS InCat 91m catamaran built at Hobart, Australia for *Buquebus* of Argentina as the CATALONIA 1 and used by *Buquebus España* on their service between Barcelona (Spain) and Mallorca. In April 2000 chartered to *P&O Portsmouth* and renamed the PORTSMOUTH EXPRESS. During winter 2000/2001 she operated for *Buquebus* between Buenos Aires (Argentina) and Piriapolis (Uruguay) and was renamed the CATALONIA. Returned to *P&O Portsmouth* in spring 2001 and was renamed the PORTSMOUTH EXPRESS. Returned to *Buquebus* in autumn 2001 and then returned to *P&O Portsmouth* in spring 2002. Laid up in Europe during winter 2002/3 and renamed the CATALONIA. She returned to *P&O Ferries* in spring 2003 trading under the marketing name 'Express'. In November she was renamed the EXPRESS. In 2004 she operated as the 'Cherbourg Express'. In 2005 to operate between Larne and Cairnryan/Troon. Additionally she is to be chartered to *Isle of Man Steam Packet* to operate some Larne – Douglas sailings during the TT period.

NORBANK Built by Van der Giessen-de Noord, Krimpen aan den IJssel, Rotterdam, Netherlands for *North Sea Ferries* for the Hull – Rotterdam service. She was originally built for and chartered to *Nedlloyd* but the charter was taken over by *P&O* in 1996 and she was bought by *P&O* in 2003. She retains Dutch crew and registry. In May 2001 moved to the Felixstowe – Europoort route. In January 2002 transferred to *P&O Irish Sea* and operated on the Liverpool – Dublin route.

NORBAY Built by Van der Giessen-de Noord, Krimpen aan den IJssel, Rotterdam, Netherlands for *North Sea Ferries* for the Hull – Rotterdam service. Owned by *P&O*. In January 2002 transferred to *P&O Irish Sea* and operated on the Liverpool – Dublin route.

Before You Go, Think...
P&O
With 3 Routes to Britain & Ireland
Larne - Cairnryan
Larne - Troon
Dublin - Liverpool
TROON
CAIRNRYAN
LARNE
DUBLIN
LIVERPOOL
For more information
see your travel agent
phone 0870 24 24 777
or log on to our website.
P&O Irish Sea
www.poirishsea.com

Pride of Le Havre *(Miles Cowsill)*

European Causeway *(Brain D.Smith)*

SEA CONTAINERS FERRIES

THE COMPANY *Sea Containers Ferries Ltd* is a British private sector company, part of the *Sea Containers Group.*

MANAGEMENT Senior Vice President, Passenger Transport David Benson.

ADDRESS Sea Containers House, 20 Upper Ground, London SE1 9PF.

TELEPHONE Administration +44 (0)20 7805 5000.

FAX Administration +44 (0)20 7805 5900.

INTERNET Email info@seacontainers.com **Website** www.seacontainers.com *(English)*

Ferry services in the UK are operated through the subsidiary *Hoverspeed Ltd. Sea Containers Ferries Scotland Ltd* (trading as *SeaCat*) has now ceased trading. See also *Silja Line* in Section 6.

HOVERSPEED

THE COMPANY *Hoverspeed Ltd* is a British private sector company. It was formed in October 1981 by the merger of *Seaspeed,* a wholly owned subsidiary of the *British Railways Board,* operating between Dover and Calais and Dover and Boulogne and *Hoverlloyd,* a subsidiary of *Broström AB* of Sweden, operating between Ramsgate (Pegwell Bay) and Calais. The Ramsgate – Calais service ceased after summer 1982. In early 1984 the company was sold by its joint owners to a management consortium. In 1986 the company was acquired by *Sea Containers.*

MANAGEMENT Managing Director Geoffrey Ede.

ADDRESS The International Hoverport, Marine Parade, DOVER, Kent CT17 9TG.

TELEPHONE Administration +44 (0)1304 865000, **Reservations** 08705 240241 (from UK only).

FAX Administration +44 (0)1304 865087, **Reservations** +44 (0)1304 240088.

INTERNET Email info@hoverspeed.co.uk **Website** www.hoverspeed.co.uk *(English)*

ROUTE OPERATED ***March-December*** Dover – Calais (55 mins; *(2,2)*; up to 9 per day). In view of interchange between routes all *Sea Containers* fast ferries (except those operated by *Silja Line* in Estonia) are shown here:

1»●	EMERAUDE FRANCE	3012t	90	35.0k	74.3m	350P	80C	-	BA	BS
2»	RAPIDE	4112t	96	37.0k	81.1m	654P	140C	-	A	LX
3»	SEACAT DIAMANT	3454t	96	37.0k	81.1m	654P	140C	-	A	IT
4»●	SEACAT SCOTLAND	3003t	91	37.0k	74.3m	450P	80C	-	BA	UK
5»	SPEEDRUNNER 1	3000t	90	37.0k	74.3m	577P	80C	-	BA	UK
6»●	THE PRINCESS ANNE	-	69	50.0k	56.4m	360P	55C	-	BA	UK
7»●	THE PRINCESS MARGARET	-	68	50.0k	56.4m	360P	55C	-	BA	UK
8»	ZARA JET	3003t	91	37.0k	74.3m	432P	80C	-	BA	PA

EMERAUDE FRANCE Built by Hobart, Tasmania as the SEACAT TASMANIA for *Sea Containers* subsidiary *Tasmanian Ferry Services* of Australia to operate between George Town (Tasmania) and Port Welshpool (Victoria). In 1992 chartered to *Hoverspeed* to operate Dover – Calais and Folkestone – Boulogne services. Returned to Australia after the 1992 summer season but returned to Britain in summer 1993 to operate Dover – Calais and Folkestone – Boulogne services during the summer. She was repainted into *Hoverspeed* livery and renamed the SEACAT CALAIS. In 1994 chartered for five years (with a purchase option) to *Navegacion Atlantida* for *Ferry Linas Argentinas AS* of Uruguay service between Montevideo (Uruguay) – Buenos Aires (Argentina) service and renamed the ATLANTIC II. The purchase option was not taken up and in 1999 she was returned to *Sea Containers* and operated for Hoverspeed between Dover and Calais. In 2000 she was chartered to *SNAV Aliscafi* of Italy to operate between Ancona (Italy) and Split (Croatia) in a joint venture with *Sea Containers* and renamed the CROAZIA JET. This operation was repeated in 2001. In 2002 renamed the SEACAT FRANCE and transferred to operate between Dover and Calais. At the end of

SeaFrance Rodin *(Miles Cowsill)*

Speed One *(John May)*

3	SEAFRANCE MANET	15093t	84	18.0k	130.0m	1800P	330C	43L	BA2	FR
4	SEAFRANCE RENOIR	15612t	81	18.0k	130.0m	1600P	330C	43L	BA2	FR
5	SEAFRANCE RODIN	34000t	01	25.0k	185.0m	1900P	700C	120L	BA2	FR

SEAFRANCE BERLIOZ Built by Chantiers de l'Atlantique, St Nazaire for *SeaFrance*.

SEAFRANCE CEZANNE Built by Kockums Varv AB, Malmö, Sweden as the ARIADNE for *Rederi AB Nordö* of Sweden. Renamed the SOCA before entering service on *UMEF* freight services (but with capacity for 175 drivers) in the Mediterranean. In 1981 she was sold to *SO Mejdunaroden Automobilen Transport (SOMAT)* of Bulgaria and renamed the TRAPEZITZA. She operated on *Medlink* services between Bulgaria and the Middle East. In 1988 she was acquired by *Sealink British Ferries*, re-registered in the Bahamas and in 1989 renamed the FANTASIA. Later in 1989 she was modified in Bremerhaven, renamed the CHANNEL SEAWAY and, in May, she inaugurated a new freight-only service between Dover (Eastern Docks) and Calais. During winter 1989/90 she was modified in Bremerhaven to convert her for passenger service. In spring 1990 she was renamed the FIESTA, transferred to *SNAT*, re-registered in France and replaced the CHAMPS ELYSEES (now the SEAFRANCE MANET) on the Dover – Calais service. In 1996 she was renamed the SEAFRANCE CEZANNE.

SEAFRANCE MANET Built by Chantiers Dubigeon SA, Nantes, France for *SNCF* as the CHAMPS ELYSEES to operate Calais – Dover and Boulogne – Dover services, later operating Calais – Dover only. In 1990 transferred to the Dieppe – Newhaven service. Chartered to *Stena Sealink Line* in June 1992 when they took over the operation of the service. She was renamed the STENA PARISIEN and carried a French crew. In 1997 the charter was terminated; she returned to *SeaFrance* and was renamed the SEAFRANCE MANET.

SEAFRANCE RENOIR Built by Ateliers et Chantiers du Havre, Le Havre, France for *SNCF* as the COTE D'AZUR for the Dover – Calais service. She also operated Boulogne – Dover in 1985. In 1996 she was renamed the SEAFRANCE RENOIR. Following the delivery of the SEAFRANCE RODIN she became the reserve vessel. During summer 2002 she operated her own roster at peak periods with reduced passenger facilities. During summer 2003 she operated a full roster to enable 20 passenger sailings per day to operate. This happened again in summer 2004. In 2005 will also operate a full roster.

SEAFRANCE RODIN Built by Aker Finnyards, Rauma, Finland for *SeaFrance*.

SMYRIL LINE

THE COMPANY *Smyril Line* is a Faroe Islands company. 20% is owned by the *Shetland Development Agency*.

MANAGEMENT Managing Director Jan Walle, **Financial Manager** Joannes A Vali.

ADDRESS Jonas Bronksgöta 37, PO Box 370, FO-100 Tórshavn, Faroe Islands.

TELEPHONE Administration +298-345900, **Reservations *Faroe Islands*** +298-345900, ***UK*** +44 (0)1595 690845 (Smyril Line Shetland).

FAX *Smyril Line* +298-343950.

INTERNET Email office@smyril-line.fo **Website** www.smyril-line.com *(English, Danish, German, Faroese, Icelandic, Norwegian)*

ROUTES OPERATED Tórshavn (Faroes) – Hanstholm (Denmark) (31 hrs; *(1)*; 1 per week), Tórshavn – Lerwick (Shetland) (12 hrs; *(1)*; 1 per week) – Bergen (Norway) (via Lerwick) (24 hrs – 27 hrs 30 mins; *(1)*; 1 per week), Tórshavn – Seydisfjordur (Iceland) (15 hrs – 18 hrs; *(1)*; 1 per week).

1	NORRÖNA	36000t	03	21.0k	164.0m	1482P	800C	134T	BA	FA

NORRÖNA Built by Flender Werft, Lübeck, Germany for *Smyril Line*, to replace the existing NORRÖNA. Originally due to enter service in summer 2002, start of building was delayed by financing difficulties. Originally to have been built at Flensburger Schiffbau-Gesellschaft, delays led to change of shipyard.

Stena Explorer *(Miles Cowsill)*

Stena Britannica *(Andrew Cooke)*

4»	STENA DISCOVERY	19638t	97	40.0k	126.6m	1500P	375C	50L	A	NL
5	STENA EUROPE	24828t	81	19.0k	149.0m	2076P	456C	60T	BA2	UK
6»	STENA EXPLORER	19638t	96	40.0k	126.6m	1500P	375C	50L	A	UK
7	STENA HOLLANDICA	33796t	01	22.0k	188.3m	380P	-	216T	BA	NL
8»	STENA LYNX III	4113t	96	35.0k	81.1m	620P	181C	-	A	UK
9»	STENA VOYAGER	19638t	96	40.0k	126.6m	1500P	375C	50L	A	UK

STENA ADVENTURER Ro-pax vessel built by Hyundai Heavy Industries, Ulsan, South Korea, for *Stena RoRo* and chartered to *Stena Line* to operate between Holyhead and Dublin.

STENA BRITANNICA Ro-pax vessel built by Hyundai Heavy Industries, Ulsan, South Korea, for *Stena RoRo*. Launched and delivered as the STENA BRITANNICA II. Chartered to *Stena Line* for use on the Hoek van Holland – Harwich service, replacing the 2000 built STENA BRITANNICA, now the FINNFELLOW of *FinnLink*. A month after delivery renamed the STENA BRITANNICA.

STENA CALEDONIA Built by Harland & Wolff Ltd, Belfast, UK for *Sealink* as the ST DAVID for the Holyhead – Dun Laoghaire and Fishguard – Rosslare services. It was originally planned that she would replace the chartered STENA NORMANDICA (5607t, 1975) but it was subsequently decided that an additional large vessel was required for the Irish Sea routes. Until 1985 her normal use was, therefore, to substitute for other Irish Sea vessels as necessary (including the Stranraer – Larne route) and also to operate additional summer services on the Holyhead – Dun Laoghaire route. During the spring of 1983 she operated on the Dover – Calais service. From March 1985 she operated between Dover and Ostend, a service which ceased in December 1985 with the decision of *RMT* to link up with *Townsend Thoresen*. During the early part of 1986 she operated between Dover and Calais and then moved to the Stranraer – Larne route (later Stranraer – Belfast) where she became a regular vessel. In 1990 she was renamed the STENA CALEDONIA. In September 1996 she became mainly a freight-only vessel but passengers were carried on certain sailings and when the STENA VOYAGER was unavailable; cars and passengers are now conveyed on all sailings.

STENA DISCOVERY Finnyards HSS1500 ('High-speed Sea Service') built at Rauma, Finland for *Stena Line* to replace two vessels on the Harwich – Hoek van Holland service.

STENA EUROPE Built by Götaverken Arendal AB, Gothenburg, Sweden as the KRONPRINSESSAN VICTORIA for *Göteborg-Frederikshavn-Linjen* of Sweden (trading as *Sessan Linjen*) for their Gothenburg – Frederikshavn service. Shortly after delivery, the company was taken over by *Stena Line* and services were marketed as *Stena-Sessan Line* for a period. In 1982 she was converted to an overnight ferry by the conversion of one vehicle deck to two additional decks of cabins and she was switched to the Gothenburg – Kiel route (with, during the summer, daytime runs from Gothenburg to Frederikshavn and Kiel to Korsør (Denmark)). In 1989 she was transferred to the Oslo – Frederikshavn route and renamed the STENA SAGA. In 1994, transferred to *Stena Line bv*, renamed the STENA EUROPE and operated between Hoek van Holland and Harwich. She was withdrawn in June 1997, transferred to the *Lion Ferry* (a *Stena Line* subsidiary) Karlskrona – Gdynia service and renamed the LION EUROPE. In 1998 she was transferred back to *Stena Line* (remaining on the same route) and renamed the STENA EUROPE. In early 2002 the cabins installed in 1982 were removed and other modifications made and she was transferred to the Fishguard – Rosslare route.

STENA EXPLORER Finnyards HSS1500 built at Rauma, Finland for *Stena RoRo* and chartered to *Stena Line*. Operates on the Holyhead – Dun Laoghaire route.

STENA HOLLANDICA Ro-pax ferry built by Astilleros Españoles, Cadiz, Spain for *Stena RoRo* and chartered to *Stena Line bv* to operate between Hoek van Holland and Harwich.

STENA LYNX III InCat 81m catamaran built at Hobart, Tasmania, Australia. Chartered new by *American Fast Ferries* of Argentina to *Stena Line* in June 1996 and named the STENA LYNX III. Initially used on the Dover – Calais service. From summer 1997 until autumn 1998 she operated between Newhaven and Dieppe. In March 1998 she was transferred to *P&O Stena Line* and renamed the ELITE. She was then renamed the P&O STENA ELITE (although only carrying the name ELITE on the bow). In late 1998 she was transferred back to *Stena Line* and renamed the STENA LYNX III. In 1999 she was placed on the Fishguard – Rosslare service, replacing the STENA LYNX (3231t,

1993). The charter ended in autumn 2000 but was immediately renewed as a summer only operation (with winter lay up). The charter ended again in autumn 2003 and she resumes the name ELITE. Later in 2003 purchased by *Stena Ropax*. In 2004 renamed the STENA LYNX III and resumed service on the Fishguard – Rosslare route, marketed as the 'Stena Express'.

STENA VOYAGER Finnyards HSS1500 built at Rauma, Finland for *Stena RoRo* and chartered to *Stena Line*. Operates on the Stranraer – Belfast route.

SUPERFAST FERRIES

THE COMPANY *Superfast Ferries* is a Greek car-passenger ferry operator, owned by the *Attica Group*.

MANAGEMENT Managing Director Alexander P Panagopulos, **Corporate Relations and Development Director/Director North Sea** Yannis B Criticos.

ADDRESS *Greece* 157, C Karamanli Avenue, Voula 16673, Greece, ***Scotland*** & ***Belgium*** ZSB, The Terminal Building, Doverlaan 7, 8380 Zeebrugge, Belgium.

TELEPHONE Administration *Greece & North Sea* +30 (0)210 891 9500, **Reservations *Scotland*** 0870 234 0870, ***Belgium*** +32 (0)50 252 252.

FAX Administration *Greece & North Sea* +30 (0)210 891 9509.

INTERNET Email criticos@attica-group.com **Website** www.superfast.com (*English, German, Dutch, French*)

ROUTES OPERATED Rosyth (Scotland) – Zeebrugge (Belgium) (17 hrs 30 mins; *(1,2)*; 1 per day). Service started in May 2002.

1	SUPERFAST IX	30441t	02	27.9k	203.3m	728P	695C	110L	BA2	GR
2	SUPERFAST X	30441t	02	27.9k	203.3m	728P	695C	110L	BA2	GR

SUPERFAST IX and SUPERFAST X Built by Howaltswerke Deutsche Werft AG, Kiel, Germany for *Attica Enterprises* (now *Attica Group*) for use by *Superfast Ferries*. SUPERFAST IX operated between Rostock and Södertälje from January until April 2002. In May 2002 both began operating between Rosyth and Zeebrugge. In 2004 fitted with additional cabins and conference/seating areas.

SWANSEA CORK FERRIES

THE COMPANY *Swansea Cork Ferries* is a company established in 1987 to re-open the Swansea – Cork service abandoned by *B&I Line* in 1979. It was originally jointly owned by *West Glamorgan County Council, Cork Corporation, Cork County Council* and *Kerry County Council*. The service did not operate in 1989 but resumed in 1990. In 1993 the company was acquired by *Strintzis Lines* of Greece. In 1999 it was purchased from *Strintzis Lines* by a consortium of Irish businessmen.

MANAGEMENT Managing Director Thomas Hunter McGowan, **Sales Manager *IR*** Simone Collins, ***UK*** John Doonican.

ADDRESS *IR* 52 South Mall, Cork, Republic of Ireland, ***UK*** Kings Docks, Swansea SA1 1SF.

TELEPHONE Administration *Head Office* +353 (0)21 427 6000, ***Cork Ferry Port*** +353 (0)21 437 8036, **Reservations *IR*** +353 (0)21 427 1166, ***UK*** +44 (0)1792 456116.

FAX Administration *IR* +353 (0)21 427 5814, ***UK*** +44 (0)1792 644356, **Reservations *IR*** +353 (0)21 427 5061, ***UK*** +44 (0)1792 644356.

INTERNET Email info@swanseacorkferries.com **Website** www.swanseacorkferries.com *(English)*

ROUTE OPERATED *March – January* Cork – Swansea (10 hrs; *(1)*; 1 per day or alternate days, according to season. Due to tidal restrictions at Swansea, the service operates to Pembroke Dock on a few days each year.

1	SUPERFERRY	15127t	72	21.0k	137.8m	1100P	300C	45L	BA	SV

SUPERFERRY Built by Hashihama Shipbuilding, Imabari City, Japan as the CASSIOPEIA for *Ocean Ferry* of Japan. She operated between Chiba (Honshu) and Tokushima (Shikoku). In 1976 the company merged with *Tokyu Ferry* and became *Ocean Tokyu Ferry;* she was renamed the IZU NO 3. She was then used on the service between Tokyo (Honshu) – Tokushima (Shikoku) – Kokura (Kyushu). In 1991 she was sold to *Strintzis Lines* and briefly renamed the IONIAN EXPRESS. Following major rebuilding, she was renamed the SUPERFERRY and used on their services between Greece and the Greek islands. In 1993 time chartered to *Swansea Cork Ferries.* The charter continued following the sale of *SCF* by *Strintzis Lines.* In winter she used to return to *Strintzis Lines* in Greece for refit and also operated for them on Mediterranean routes. The charter was not renewed for 2001 and she remained in the Mediterranean, operating for *Strintzis Lines* under the *Blue Ferries* name (following the reorganisation of the company after its acquisition by *Attica Enterprises*). She was renamed the BLUE AEGEAN. In 2002 she was sold to *Swansea Cork Ferries* and renamed the SUPERFERRY.

TRANSEUROPA FERRIES

THE COMPANY *TransEuropa Ferries NV* is a Belgian subsidiary of *TransEuropa Shipping Lines,* a Slovenian private sector company. Channel operations started in 1997, in conjunction with *Sally Ferries,* replacing them on November 1998. Passenger operations started in 2004. The company traded as *TransEuropa Shipping Lines (TSL)* until 2000. Note that all the owning companies listed here are associated companies of *TSL.*

MANAGEMENT *TransEuropa Shipping* Managing Director Stergulc Rihard, ***TransEuropa Ferries NV*, General Manager Belgium & UK** Mr Dominique Penel, **Sales Manger, Europe** Mr Peter Sys.

ADDRESS *TSL Slovenia* Vojkovo nabrezje 38, 6000 Koper, Slovenia, ***TEF UK*** Ferry Terminal, Ramsgate New Port, RAMSGATE, Kent CT11 8RP ***TEF Belgium*** Slijkensesteenweg 2, 8400 Ostend, Belgium.

TELEPHONE Admin *TSL Slovenia* +386 (0)5 664 17 77, ***TEF UK*** +44 (0)1843 853833, ***TEF Belgium*** +32 (0)59 34 02 60. **Reservations *TEF UK*** 01843 595522.

FAX *TSL Slovenia* +386 (0)5 639 50 36, ***TEF UK*** +44 (0)1843 594663, ***TEF Belgium*** +32 (0)59 34 02 61.

INTERNET Website www.transeuropaferries.com *(English)*

ROUTE OPERATED Ramsgate – Ostend (Belgium) (4 hrs; *(1,2,3)*; up to 4 per day).

1	EUROVOYAGER	12110t	78	22.0k	118.4m	1250P	348C	45L	BA2	CY
2	LARKSPUR	14458t	76	17.5k	143.8m	1150P	314C	55L	BA2	BS
3	OLEANDER	13728t	80	23.0k	132.5m	1326P	350C	44L	BA2	CY
4	WISTERIA	13505t	78	19.0k	131.0m	1500P	354C	38T	BA	CY

EUROVOYAGER Built by NV Cockerill Yards, Hoboken, Belgium as the PRINS ALBERT for RMT of Belgium for the Ostend – Dover service. During 1986 she had an additional vehicle deck added. In 1994 the British port became Ramsgate. Withdrawn after 28th February 1997 and laid up. In 1998 she was sold to *Hawthorn Shipping Co Ltd,* renamed the EUROVOYAGER. In July, she was chartered to *Sally Freight.* In November the *Sally Freight* service ended and she immediately began operating for *TSL* with no break in service. From 2004 began operating in passenger mode when the LARKSPUR or OLEANDER were unavailable.

LARKSPUR Built by Schichau-Unterweser AG, Bremerhaven, Germany as the GEDSER for *Gedser-Travemünde Ruten* of Denmark for their service between Gedser (Denmark) and Travemünde (Germany). In 1986 she was purchased by *Thorsviks Rederi A/S* of Norway and chartered to *Sally Ferries,* re-registered in the Bahamas, renamed the VIKING 2 and entered service on the Ramsgate – Dunkerque service. In early 1989 she was renamed the SALLY SKY and during winter 1989/90 she was 'stretched' to increase vehicle capacity. At the end of 1996 she was withdrawn from the Dunkerque service. In 1997 she was renamed the EUROTRAVELLER, transferred to *Holyman-Sally Ferries* and, in March, was introduced onto the Ramsgate – Ostend route. In 1998, when *Holyman-*

Larkspur *(Miles Cowsill)*

Dieppe *(FotoFlite)*

Sally Ferries came to an end, she operated in a freight-only role for *Sally Line* under the *Sally Freight* name. Passenger services were resumed in May, under the name of *Sally Direct*. All *Sally Line* operations ended in November 1998 and she was withdrawn for sale and laid up. In 1999 sold to *Forsythia Maritime Co Ltd* and renamed the LARKSPUR. She was given a major refit at Dunkerque, including the provision of 60 drivers' cabins with private facilities. She entered service with *TEF* in August 2000. Passengers were conveyed from July 2004.

OLEANDER Built by Schichau-Unterweser AG, Bremerhaven, Germany for *European Ferries (Townsend Thoresen)* as the PRIDE OF FREE ENTERPRISE for the Dover – Calais service, also operating on the Dover – Zeebrugge service during the winter. In 1988 she was renamed the PRIDE OF BRUGES and, following the delivery of the new PRIDE OF CALAIS, she was transferred all year to the Dover – Zeebrugge service. In 1992, after the closure of that routes to passengers, she returned to the Dover – Calais route. Plans to operate her in a freight-only mode in 1997 were changed and she ran as a full passenger vessel. In 1998, transferred to *P&O Stena Line*; plans to transfer her to the Newhaven – Dieppe route were dropped and she remained at Dover. In 1999 renamed the P&OSL PICARDY. In early 2000 she was laid up for sale in Dunkerque. In 2001 she was sold to *Seaborne Navigation Co Ltd* and renamed the OLEANDER. Entered service with *TEF* in July 2002 after major renovation work in Dunkerque, including the provision of 60 drivers' cabins with private facilities. Passengers were conveyed from July 2004.

WISTERIA Built at Heuseden, Netherlands as the PRINSES BEATRIX for *Stoomvaart Maatschappij Zeeland (Zeeland Steamship Company)* of The Netherlands for their Hoek van Holland – Harwich service. In September 1985 sold to *Brittany Ferries* and chartered back to *SMZ*, continuing to operate for them until the introduction of the KONINGIN BEATRIX (see STENA BALTICA) in May 1986. In June 1986 delivered to *Brittany Ferries*, renamed the DUC DE NORMANDIE and inaugurated the Portsmouth – Caen service. From July 2002 she moved to the Plymouth – Roscoff route replacing the QUIBERON (11813t, 1975). Withdrawn at the end of the 2004 summer season and laid up at Gdansk. In 2005 sold to as associated company of *TransEuropa Shipping* and moved briefly to Ostend. Chartered to *Ferrimaroc* to operate between Almeria (Spain) and Nador (Morocco). May return to Ostend in due course.

TRANSMANCHE FERRIES

Transmanche Ferries is a French company, controlled by the *Syndicat Mixte de L'Activité Transmanche*.

ADDRESS Transmanche Ferries, Quai Gaston Lalitte, 76200 Dieppe, France.

TELEPHONE Administration +33 02 32 14 52 03, **Reservations *UK*** 0800 917 1201, ***France*** 0800 650 100.

FAX Administration +33 02 32 14 52 00.

INTERNET Website www.transmancheferries.com *(English, French)*

ROUTE OPERATED Newhaven – Dieppe (4 hrs; *(1,2)*; 4 per day).

1	DIEPPE	17672t	81	19.0k	145.9m	250P	-	140T	BA	FR
2	SARDINIA VERA	11637t	75	18.5k	120.8m	700P	479C	58T	BA	IT

DIEPPE Built by Fartygsentreprenader AB, Kalmar, Sweden as the SAGA STAR for *TT-Saga-Line* and, from 1982, used on freight services between Travemünde and Trelleborg/Malmö. (Originally ordered by *Rederi AB Svea* as the SAGALAND). In 1989 sold to *Cie Meridonale* of France, renamed the GIROLATA and used on *SNCM* (later *CMR*) services in the Mediterranean. In 1993 she was chartered back to *TT-Line*, resumed her original name and was used on the Travemünde – Trelleborg service. Following delivery of the ROBIN HOOD and the NILS DACKE in 1995, she was transferred to the Rostock – Trelleborg route. In July 1997 she was purchased by *TT-Line* and in 1998 passenger facilities were completely renovated to full ro-pax format; following the delivery of the TOM SAWYER she was transferred back to the Travemünde – Trelleborg route, operating additional freight sailings. Briefly transferred back to Rostock – Trelleborg when the charter of the TT-TRAVELLER ended. Withdrawn in 2002, sold to *Transmanche Ferries* and renamed the DIEPPE.

SARDINIA VERA Built by Rickmers Werft, Bremerhaven, Germany for *Stena Line* of Sweden. Laid down as the STENA ATLANTICA but launched as the MARINE ATLANTICA. On delivery, chartered to *CN Marine* (from 1986 *Marine Atlantic*) of Canada for service between North Sydney (Nova Scotia) and Port-aux-Basques (Newfoundland). In 1986 she was sold to *Tourship Co AS* of Italy (parent company of *Corsica Ferries*) and, in 1987, renamed the CORSICA VERA. Later in 1987 renamed the SARDINIA VERA. Used on services between Italy and France and Corsica and Sardinia. In 2001 chartered to the *Transmanche Ferries*.

Under Construction

3	NEWBUILDING 1	-	05	22.0k	112.0m	600P	-	62L	BA	FR
4	NEWBUILDING 2	-	06	22.0k	112.0m	600P	-	62L	BA	FR

NEWBUILDING 1, NEWBUILDING 2 Under construction by Astilleros Barreras SA, Vigo, Spain to replace the DIEPPE and SARDINIA VERA. First due to be delivered October 2005.

Superfast IX & **Pride of Bruges** *(Miles Cowsill)*

RED FUNNEL

Red Osprey & **Red Jet 4** *(Andrew Cooke)*

section 2

domestic services

ARGYLL AND BUTE COUNCIL

THE COMPANY *Argyll and Bute Council* is a British local government authority.

MANAGEMENT Director of Operations Andrew Law, **Head of Roads & Amenity** Stewart Turner, **Assistant Operations Manager** Martin Gorringe.

ADDRESS Manse Brae, Lochgilphead, Argyll PA31 8RD.

TELEPHONE Administration +44 (0)1546 604614.

FAX Administration +44 (0)1546 606443.

INTERNET Email martin.gorringe@argyll-bute.gov.uk **Website** www.argyll-bute.gov.uk *(English)*

ROUTES OPERATED Vehicle ferries Seil – Luing (5 mins; *(1)*; approx half hourly), Port Askaig (Islay) – Feolin (Jura) (5 mins; *(3)*; approx hourly). **Passenger only ferries** Port Appin – Lismore (10 mins; *(4)*; approx half hourly), Ellenabeich – Easdale (5 mins; *(2)*; as required).

1	BELNAHUA	35t	72	8.0k	17.1m	40P	5C	1L	BA	UK
2p	EASDALE	-	93	6.5k	6.4m	11P	0C	0L	-	UK
3	EILEAN DHIURA	86t	98	9.0k	25.6m	50P	13C	1L	BA	UK
4p	LISMORE	12t	88	8.0k	9.7m	20P	0C	0L	-	UK

BELNAHUA Built by Campbeltown Shipyard, Campbeltown, UK for *Argyll County Council* for the Seil – Luing service. In 1975, following local government reorganisation, transferred to *Strathclyde Regional Council*. In 1996, transferred to *Argyll and Bute Council*.

EASDALE Built for *Strathclyde Regional Council* for the Ellenabeich – Easdale passenger only service. In 1996, following local government reorganisation, transferred to *Argyll and Bute Council*.

EILEAN DHIURA Built by McTay Marine, Bromborough, Wirral, UK for *Argyll and Bute County Council* to replace the *Western Ferries (Argyll)* SOUND OF GIGHA on the Islay – Jura route. *ASP Seascot* manage and operate this vessel on behalf of the *Argyll and Bute Council*.

LISMORE Built for *Strathclyde Regional Council* for the Port Appin – Lismore passenger only service. In 1996, following local government reorganisation, transferred to *Argyll and Bute Council*. *Foster Yeoman* manage and operate this vessel on behalf of the *Argyll and Bute Council*.

ARRANMORE ISLAND FERRY SERVICES

THE COMPANY *Arranmore Island Ferry Services* is an Irish Republic company, supported by *Údarás na Gaeltachta (The Gaeltacht Authority)*, a semi-state owned body responsible for tourism and development in the Irish speaking areas of The Irish Republic. The operation is also known as *Maoin-Na-Farraige* (literally 'sea treasure' or 'sea wealth').

MANAGEMENT Managing Director Cornelius Bonner.

ADDRESS Bridge House, Leabgarrow, Arranmore, County Donegal, Republic of Ireland.

TELEPHONE Administration & Reservations +353 (0)7495 20532.

FAX Administration & Reservations + 353 (0)7495 20750.

Clansman *(CalMac)*

Coruisk *(CalMac)*

3	CALEDONIAN ISLES	52211t	93	15.0k	94.3m	1000P	120C	10L	BA	UK
4	CANNA	69t	76	8.0k	24.3m	140P	6C	-	B	UK
5	CLANSMAN	5499t	98	16.5k	99.0m	638P	90C	6L	BA	UK
6	CORUISK	1108t	03	14.0k	65.0m	250P	40C	-	BA	UK
7	EIGG	69t	75	8.0k	24.3m	75P	6C	-	B	UK
8	HEBRIDEAN ISLES	3040t	85	15.0k	85.1m	494P	68C	10L	BAS	UK
9	HEBRIDES	5506t	00	16.5k	99.0m	612P	110C	6L	BA	UK
10	ISLE OF ARRAN	3296t	84	15.0k	85.0m	446P	68C	8L	BA	UK
11	ISLE OF CUMBRAE	201t	77	8.5k	37.7m	139P	18C	-	BA	UK
12	ISLE OF LEWIS	6753t	95	18.0k	101.2m	680P	123C	10L	BA	UK
13	ISLE OF MULL	4719t	88	15.0k	90.1m	962P	80C	20L	BA	UK
14	JUNO	854t	74	13.0k	69.0m	381P	40C	-	AS	UK
15	JUPITER	848t	74	13.0k	69.0m	381P	40C	-	AS	UK
16	LOCH ALAINN	396t	98	10.0k	43.0m	150P	24C	-	BA	UK
17	LOCH BHRUSDA	246t	96	8.0k	35.4m	150P	18C	-	BA	UK
18	LOCH BUIE	295t	92	9.0k	35.5m	250P	9C	-	BA	UK
19	LOCH DUNVEGAN	549t	91	9.0k	54.2m	200P	36C	-	BA	UK
20	LOCH FYNE	549t	91	9.0k	54.2m	200P	36C	-	BA	UK
21	LOCH LINNHE	206t	86	9.0k	35.5m	199P	12C	-	BA	UK
22	LOCH PORTAIN	950t	03	10.5k	50.0m	200P	32C	-	BA	UK
23	LOCH RANZA	206t	87	9.0k	35.7m	199P	12C	-	BA	UK
24	LOCH RIDDON	206t	86	9.0k	35.5m	199P	12C	-	BA	UK
25	LOCH STRIVEN	206t	86	9.0k	35.7m	199P	12C	-	BA	UK
26	LOCH TARBERT	211t	92	9.0k	34.5m	149P	18C	-	BA	UK
27	LOCHNEVIS	941t	00	13.0k	49.1m	190P	14C	-	A	UK
28	LORD OF THE ISLES	3504t	89	16.0k	84.6m	506P	56C	16L	BAS	UK
29	RAASAY	69t	76	8.0k	24.3m	75P	6C	-	B	UK
30	SATURN	851t	78	13.0k	69.5m	381P	40C	-	AS	UK

ALI CAT Catamaran built for *Solent & White Line Cruises* of Ryde, Isle of Wight. Operated a passenger service between Cowes to Hamble and Warsash and cruises from Cowes. At times chartered to *Wightlink* to cover for the fast catamarans when there were only two in their fleet. In 2002 chartered to *Red Funnel Ferries* who had contracted with *Caledonian MacBrayne* to operate passenger only services between Gourock and Dunoon in the morning and evening peaks.

BRUERNISH Built by James Lamont & Co, Port Glasgow, UK. Until 1980 she served on a variety of routes. In 1980 she inaugurated ro-ro working between Tayinloan and the island of Gigha and served this route until June 1992 when she was replaced by the LOCH RANZA and became a relief vessel. In summer 1994 she operated as secondary vessel on the Tobermory (Mull) – Kilchoan service for one season only. In December 1996 she started a vehicle ferry service between Ballycastle (on the North West coast of Northern Ireland) and Rathlin Island under charter; the route became a *Caledonian MacBrayne* operation in April 1997 – see the CANNA. In 1997 she operated on the Tarbert – Portavadie service and in 1998 on the Oban – Lismore service. Since 1999 she has been a spare vessel.

CALEDONIAN ISLES Built by Richards Shipyard, Lowestoft, UK for the Ardrossan – Brodick (Arran) service.

CANNA Built by James Lamont & Co Ltd, Port Glasgow, UK. She was the regular vessel on the Lochaline – Fishnish (Mull) service. In 1986 she was replaced by the ISLE OF CUMBRAE and until 1990 she served in a relief capacity in the north, often assisting on the Iona service. In 1990 she replaced the KILBRANNAN (see the ÁRAINN MHÓR, *Arranmore Island Ferry Services*) on the Kyles Scalpay (Harris) – Scalpay service (replaced by a bridge in autumn 1997). In spring 1997 she was transferred to the Ballycastle – Rathlin Island route.

CLANSMAN Built by Appledore, UK to replace the LORD OF THE ISLES on the Oban – Coll and Tiree

Jupiter *(Miles Cowsill)*

Eynhallow *(Miles Cowsill)*

and Oban – Castlebay and Lochboisdale service in the summer. She also serves as winter relief vessel on the Stornoway, Tarbert, Lochmaddy, Mull/Colonsay, Islay and Brodick routes.

CORUISK Built by Appledore Shipbuilders Ltd, Appledore, UK to replace the LORD OF THE ISLES on the Mallaig – Armadale route during the summer. She operates on the Upper Clyde during the winter.

EIGG Built by James Lamont & Co, Port Glasgow, UK. Since 1976 she was employed mainly on the Oban – Lismore service. In 1996 she was transferred to the Tobermory (Mull) – Kilchoan route, very occasionally making special sailings to the Small Isles (Canna, Eigg, Muck and Rum) for special cargoes. In 1999 her wheelhouse was raised to make it easier to see over taller lorries and she returned to the Oban – Lismore route.

HEBRIDEAN ISLES Built by Cochrane Shipbuilders, Selby UK for the Uig – Tarbert/Lochmaddy service. She was used initially on the Ullapool – Stornoway and Oban – Craignure/Colonsay services pending installation of link-span facilities at Uig, Tarbert and Lochmaddy. She took up her regular role in May 1986. Since May 1996 she no longer operated direct services in summer between Tarbert and Lochmaddy, this role being taken on by the new Harris – North Uist services of the LOCH BHRUSDA. In 2001 replaced by the HEBRIDES and transferred to the Islay service. In autumn 2002 she operated between Scrabster and Stromness for *NorthLink Orkney and Shetland Ferries* before port modifications at Scrabster enabled the HAMNAVOE to enter service in spring 2003. She then returned to the Islay service. She also relieved on the Pentland Firth in 2004.

HEBRIDES Built by Ferguson Shipbuilders Ltd, Port Glasgow, UK for the Uig – Tarbert and Uig – Lochmaddy services.

ISLE OF ARRAN Built by Ferguson Ailsa, Port Glasgow, UK for the Ardrossan – Brodick service. In 1993 transferred to the Kennacraig – Port Ellen/Port Askaig service, also undertaking the weekly Port Askaig – Colonsay – Oban summer service. From then until 1997/98 she also relieved on the Brodick, Coll/Tiree, Castlebay/Lochboisdale, Craignure and Tarbert/Lochmaddy routes in winter. In 2001 replaced by the HEBRIDEAN ISLES and is now reserve for the larger vessels. She has operated on the two-ship Islay service in summer since 2003.

ISLE OF CUMBRAE Built by Ailsa Shipbuilding Ltd, Troon, UK for the Largs – Cumbrae Slip (Cumbrae) service. In 1986 she was replaced by the LOCH LINNHE and the LOCH STRIVEN and transferred to the Lochaline – Fishnish (Mull) service. She used to spend most of the winter as secondary vessel on the Kyle of Lochalsh – Kyleakin service; however this ceased following the opening of the Skye Bridge in 1995. In 1997 she was transferred to the Colintraive – Rhubodach service. In summer 1999 she was transferred to the Tarbert – Portavadie service.

ISLE OF LEWIS Built by Ferguson Shipbuilders Ltd, Port Glasgow, UK for the Ullapool – Stornoway service.

ISLE OF MULL Built by Appledore Ferguson, Port Glasgow, UK for the Oban – Craignure (Mull) service. She also operates the Oban – Colonsay service and until 1997/98 was the usual winter relief vessel on the Ullapool – Stornoway service. She has also deputised on the Oban – Castlebay/Lochboisdale and Oban – Coll/Tiree routes.

JUNO, JUPITER, SATURN Built by James Lamont, Port Glasgow, UK (SATURN Ailsa Shipbuilding, Troon) for the Gourock – Dunoon, Gourock – Kilcreggan and Wemyss Bay – Rothesay services. The JUPITER has been upgraded for a while to Class III standard for the Ardrossan – Brodick service. Before 1986, the JUNO and JUPITER operated mainly on the Gourock – Dunoon and Gourock – Kilcreggan (now withdrawn) services and the SATURN on the Wemyss Bay – Rothesay service. Since 1986 they have usually rotated on these services; until 2000 this, in summer, included Clyde cruising but this was not repeated in 2001. In summer 2005, following the delivery of the new BUTE, the SATURN will operate additional sailings between Ardrossan and Brodick with a maximum capacity of 250 passengers.

LOCH ALAINN Built by Buckie Shipbuilders Ltd, Buckie, UK for the Lochaline – Fishnish service. Launched as the LOCH ALINE but renamed the LOCH ALAINN before entering service. After a brief period on the service she was built for, she was transferred to the Colintraive – Rhubodach route. In summer 1998 she was transferred to the Largs – Cumbrae Slip service.

LOCH BHRUSDA Built by McTay Marine, Bromborough, Wirral, UK to inaugurate a new Otternish (North Uist) – Leverburgh (Harris) service. In 2001 the service became Berneray – Leverburgh. In 2003 moved to the Eriskay – Barra service, previously operated by *Comhairle Nan Eilean Siar* vessels. Note 'Bhrusda' is pronounced "Vroosta".

LOCH BUIE Built by J W Miller & Sons Ltd, St Monans, UK for the Fionnphort (Mull) – Iona service to replace the MORVERN (see *Arranmore Island Ferry Services*) and obviate the need for a relief vessel in the summer. Due to height restrictions, loading arrangements for vehicles taller than private cars are bow only. Only islanders' cars and service vehicles (eg mail vans, police) are carried; no tourist vehicles are conveyed.

LOCH DUNVEGAN Built by Ferguson Shipbuilders Ltd, Port Glasgow, UK for the Kyle of Lochalsh – Kyleakin service. On the opening of the Skye Bridge in October 1995 she was withdrawn from service and put up for sale. In autumn 1997, returned to service on the Lochaline – Fishnish route. In 1998 she was due to be transferred to the Colintraive – Rhubodach route but this was delayed because of problems in providing terminal facilities. She operated on the Clyde and between Mallaig and Armadale during the early summer and spent the rest of the summer laid up. In 1999 she was transferred to the Colintraive – Rhubodach route.

LOCH FYNE Built by Ferguson Shipbuilders Ltd, Port Glasgow UK for the Kyle of Lochalsh – Kyleakin service (see the LOCH DUNVEGAN). In autumn 1997, she also served on the Lochaline – Fishnish route and was transferred to this route as regular vessel in 1998.

LOCH LINNHE Built by Richard Dunston (Hessle) Ltd, Hessle, UK. Until 1997 she was used mainly on the Largs – Cumbrae Slip (Cumbrae) service and until winter 1994/95 she was usually used on the Lochaline – Fishnish service during the winter. Since then she had relieved on various routes in winter. In summer 1998 she operated mainly on the Tarbert – Portavadie route. In 1999 she was transferred to the summer only Tobermory – Kilchoan service. In 2003 she launched the new Sound of Barra before delivery of the LOCH PORTAIN allowed the LOCH BHRUSDA to be moved to that route.

LOCH PORTAIN Built by McTay Marine, Bromborough, Wirral, UK (hull constructed in Poland) to replace the LOCH BHRUSDA on the Berneray – Leverburgh service.

LOCH RANZA Built by Richard Dunston (Hessle) Ltd, Hessle, UK for the Claonaig – Lochranza (Arran) seasonal service and used a relief vessel in the winter. In 1992 she was replaced by the LOCH TARBERT and transferred to the Tayinloan – Gigha service.

LOCH RIDDON Built by Richard Dunston (Hessle) Ltd, Hessle, UK. Until 1997 she was used almost exclusively on the Colintraive – Rhubodach service. In 1997, she was transferred to the Largs – Cumbrae Slip service and is often to be found on the Tarbert-Portavadie/Lochranza service in winter.

LOCH STRIVEN Built by Richard Dunston (Hessle) Ltd, Hessle, UK. Used mainly on the Largs – Cumbrae Slip service until 1997. In winter 1995/6 and 1996/67 she was used on the Tarbert – Portavadie and Claonaig – Lochranza routes. In 1997 she took over the Sconser – Raasay service.

LOCH TARBERT Built by JW Miller & Sons Ltd, St Monans, UK for the Claonaig – Lochranza service. She has been the winter relief vessel on the Largs – Cumbrae Slip route since winter 1994/5 and is also the winter relief vessel for the Otternish – Leverburgh route.

LOCHNEVIS Built by Ailsa Shipbuilding, Troon, UK to replace the LOCHMOR on the Mallaig – Small Isles service and the winter Mallaig – Armadale service. Although a vehicle ferry, cars are not normally carried to the Small Isles; the ro-ro facility is used for the carriage of agricultural machinery and livestock and it is possible to convey a vehicle on the ferry from which goods can be unloaded directly on to local transport rather than transhipping at Mallaig.

LORD OF THE ISLES Built by Appledore Ferguson, Port Glasgow, UK to replace the CLAYMORE on the Oban – Castlebay and Lochboisdale services and also the COLUMBA (1420t, 1964) on the Oban – Coll and Tiree service. She took over Mallaig – Armadale and Mallaig – Outer Isles service in July 1998 but returned to her previous routes during the winter period. In spring 2003 the Mallaig – Armadale service was taken over by the PIONEER standing in for the new CORUISK and she operated services from Oban to South Uist and Barra.

RAASAY Built by James Lamont & Co Ltd, Port Glasgow, UK for and used primarily on the Sconser (Skye) – Raasay service. In 1997 she was replaced by the LOCH STRIVEN and became a spare/relief vessel and inaugurated in October 2003 the winter service between Tobermory (Mull) and Kilchoan (Ardnamurchan).

Caledonian MacBrayne also operates the LAIG BAY, an 8k, 10.5m motor vessel built in 2000 and carrying up to 28 passengers. She initially tendered to the LOCHNEVIS at Eigg. Following the building of a ramp on the island in 2004, she is now a crew tender at Tobermory where the ship is moored offshore at night.

Under Construction

31	BUTE	2300t	05	13.0k	69.0m	450P	60C	-	AS	UK

BUTE Under construction by Stocznia Remontowa, Gdansk, Poland to operate on the Wemyss Bay – Rothesay route. Enters service summer 2005.

To be ordered

32	NEWBUILDING 1	c2300t	06	13.0k	69.0m	450P	60C	-	AS	UK
33	NEWBUILDING 2	-	06	13.0k	-	-	-	-	BA	UK

NEWBUILDING 1 To be ordered during 2005 for the Wemyss Bay – Rothesay route.

NEWBUILDING 2 To be ordered during 2005 for the Largs – Cumbrae Slip route.

CROMARTY FERRY COMPANY

THE COMPANY The *Cromarty Ferry Company* operate under contract to *The Highland Council.*

MANAGEMENT Managing Director Mr John Henderson.

ADDRESS Udale Farm, Poyntzfield, BY DINGWALL, IV7 8LY.

TELEPHONE +44 (0)1381 610269, Mobile +44 (0)7768 653674.

FAX +44 (0)1381 610408.

INTERNET Email john@udalefarm.com **Website** www.cromarty-ferry.co.uk *(English)*

ROUTE OPERATED ***June-October*** Cromarty – Nigg (Ross-shire) (10 mins; *(1)*; half hourly – 0800 to 1800 ex Cromarty (19.00 in July and August)).

1	CROMARTY ROSE	28t	87	8.0k	14.0m	50P	2C	-	B	UK

CROMARTY ROSE Built by McGrindle Shipbuilding Ltd, Ardrossan, UK for *Seaboard Marine (Nigg) Ltd* who operated the service until 2001, supported by *The Highland Council.* In 2002, after a tendering exercise held in 2001, the contract was awarded to the *Cromarty Ferry Company.* The new company purchased the CROMARY ROSE from *Seaboard Marine (Nigg) Ltd.*

CROSS RIVER FERRIES

THE COMPANY *Cross River Ferries Ltd* is an Irish Republic company, jointly owned by *Marine Transport Services Ltd* of Cobh and *Arklow Shipping Ltd* of Arklow, County Wicklow.

MANAGEMENT Operations Manager Eoin O'Sullivan.

ADDRESS Westlands House, Rushbrooke, Cobh, County Cork, Republic of Ireland.

TELEPHONE Administration +353 (0)21 481 1223, **Reservations** Not applicable.

FAX Administration +353 (0)21 481 2645, **Reservations** Not applicable.

ROUTE OPERATED Carrigaloe (near Cobh, on Great Island) – Glenbrook (Co Cork) (4 mins; *(1,2)*; frequent service 07.00 – 00.15 (one or two vessels used according to demand)).

1	CARRIGALOE	‡225t	70	8.0k	49.1m	200P	27C	-	BA	IR

2	GLENBROOK	‡225t	71	8.0k	49.1m	200P	27C	-	BA	IR

CARRIGALOE Built by Newport Shipbuilding and Engineering Company, Newport (Gwent) UK as the KYLEAKIN for *Caledonian Steam Packet Company* (later *Caledonian MacBrayne*) for the Kyle of Lochalsh – Kyleakin service. In 1991 sold to *Marine Transport Services Ltd* and renamed them the CARRIGALOE. She entered service in March 1993. In summer 2002 chartered to the *Lough Foyle Ferry Company*, returning in spring 2003.

GLENBROOK Built by Newport Shipbuilding and Engineering Company, Newport (Gwent) UK as the LOCHALSH for *Caledonian Steam Packet Company* (later *Caledonian MacBrayne*) for the Kyle of Lochalsh – Kyleakin service. In 1991 sold to *Marine Transport Services Ltd* and renamed the GLENBROOK. She entered service in March 1993.

GLENELG – KYLERHEA FERRY

THE COMPANY The *Glenelg – Kylerhea Ferry* is privately operated. The company is currently for sale.

MANAGEMENT Ferry Master R MacLeod (liable to change).

ADDRESS Corriehallie, Inverinate, Kyle IV40 8HD.

TELEPHONE Administration & Reservations +44 (0)1599 511302.

FAX Administration & Reservations +44 (0)1599 511477.

INTERNET Email roddy@skyeferry.co.uk **Website** www.skyeferry.co.uk *(English)*

ROUTE OPERATED ***Easter – October only*** Glenelg – Kylerhea (Skye) (10 mins; *(1)*; frequent service).

1	GLENACHULISH	44t	69	9.0k	20.0m	12P	6C	-	BSt	UK

GLENACHULISH Built by Ailsa Shipbuilding Company, Troon, UK for the *Ballachulish Ferry Company* for the service between North Ballachulish and South Ballachulish, across the mouth of Loch Leven. In 1975 the ferry was replaced by a bridge and she was sold to *Highland Regional Council* and used on a relief basis on the North Kessock – South Kessock and Kylesku – Kylestrome routes. In 1984 she was sold to the operator of the Glenelg – Kylerhea service. She is the last turntable ferry in operation.

THE HIGHLAND COUNCIL

THE COMPANY *The Highland Council* (previously *Highland Regional Council*) is a British local government authority.

MANAGEMENT Area Transport, Environment & Community Works Services Manager James C Tolmie, **Ferry Manager** J McAuslane.

ADDRESS ***Area Office*** Lochybridge Depot, Carr's Corner Industrial Estate, Fort William PH33 6TQ, ***Ferry Office*** Ferry Cottage, Ardgour, Fort William.

TELEPHONE Administration ***Area Office*** +44 (0)1397 709000, ***Corran*** +44 (0)1855 841243, ***Camusnagaul*** +44 (0)1397 772483, **Reservations** Not applicable.

FAX Administration ***Area Office*** +44 (0)1397 705735, ***Corran*** +44 (0)1855 841243, **Reservations** Not applicable.

INTERNET Email tecs@highland.gov.uk **Website** lochabertransport.org.uk/corranferry.html *(English – external site reproducing official ferry leaflet)*

ROUTES OPERATED Vehicle Ferries Corran – Ardgour (5 mins; *(2,3)*; half hourly), **Passenger Only Ferry** Fort William – Camusnagaul (10 mins; *(3)*; Frequent).

1p	CAILIN AN AISEAG	-	80	7.5k	9.8m	26P	0C	0L	-	UK
2	CORRAN	351t	01	10.0k	42.0m	150P	30C	2L	BA	UK

Hrossey *(Miles Cowsill)*

Geira *(Miles Cowsill)*

3	MAID OF GLENCOUL	‡166t	75	8.0k	32.0m	116P	16C	1L	BA	UK

CAILIN AN AISEAG Built by Buckie Shipbuilders Ltd, Buckie, UK for *Highland Regional Council* and used on the Fort William – Camusnagaul service.

CORRAN Built by George Prior Engineering Ltd, Hull, UK for *The Highland Council* to replace the MAID OF GLENCOUL as main vessel.

MAID OF GLENCOUL Built by William McCrindle Ltd, Shipbuilders, Ardrossan, UK for *Highland Regional Council* for the service between Kylesku and Kylestrome. In 1984 the ferry service was replaced by a bridge and she was transferred to the Corran – Ardgour service. In April 1996, ownership transferred to *The Highland Council*. In 2001 became the reserve vessel.

ISLES OF SCILLY STEAMSHIP COMPANY

THE COMPANY *Isles of Scilly Steamship Company* is a British private sector company.

MANAGEMENT Chief Executive J Marston, **Marketing Manager** L Ramzan.

ADDRESS *Scilly* PO Box 10, Hugh Town, St Mary's, Isles of Scilly TR21 0LJ, ***Penzance*** Steamship House, Quay Street, Penzance, Cornwall, TR18 4BZ.

TELEPHONE Administration & Reservations *Scilly* +44 (0)1720 424220, ***Penzance*** +44 (0)1736 334220.

FAX Administration & Reservations *Scilly* +44 (0)1720 422192, ***Penzance*** +44 (0)1736 351223.

INTERNET Email sales@islesofscilly-travel.co.uk **Website** www.ios-travel.co.uk *(English)*

ROUTES OPERATED Penzance – St Mary's (Isles of Scilly) (2 hrs 40 mins; *(1,3)*; 1 per day), St Mary's – Tresco/St Martin's/St Agnes/Bryher; *(2)*; irregular).

1	GRY MARITHA	590t	81	10.5k	40.3m	6P	5C	1L	C	UK
2	LYONESSE LADY	40t	91	9.0k	15.5m	12P	1C	0L	AC	UK
3	SCILLONIAN III	1256t	77	15.5k	67.7m	600P	5C	-	C	UK

GRY MARITHA Built by Moen Slip AS, Kolvereid, Norway for *Gjofor* of Norway. In design she is a coaster rather than a ferry. In 1990 sold to *Isles of Scilly Steamship Company*. She operates a freight and passenger service all year (conveying most goods to and from the Islands). During the winter she provides the only sea service to the islands, the SCILLONIAN III being laid up.

LYONESSE LADY Built by at Fort William, UK, for inter-island ferry work.

SCILLONIAN III Built by Appledore Shipbuilders, Appledore, UK for the Penzance-St Mary's service. She operates from late March to November and is laid up in the winter. Last major conventional passenger/cargo ferry built for UK waters and probably Western Europe. Extensively refurbished during winter 98/99. She can carry cars in her hold and on deck, as well as general cargo/perishables, boats, trailer tents and passenger luggage.

KERRERA FERRY

THE COMPANY The *Kerrera Ferry* is privately operated.

MANAGEMENT Ferry Master Duncan MacEachen.

ADDRESS The Ferry, Isle of Kerrera, By Oban PA34 4SX.

TELEPHONE Administration +44 (0)1631 563665.

ROUTE OPERATED Gallanach (Argyll) – Kerrera (5 mins; *(1)*; on demand 10.30 – 12.30 and 14.00 – 18.00, Easter – October, other times by arrangement).

1	GYLEN LADY	9t	99	8.0k	10.0m	12P	1C	-	B	UK

GYLEN LADY Built by Corpach Boatyard, Corpach, UK to inaugurate a vehicle ferry service to the Isle of Kerrera, replacing open passenger boat.

Claymore *(Miles Cowsill)*

Red Jet 1 *(Andrew Cooke)*

GOLDEN MARIANA Built by Bideford Shipyard Ltd, Bideford, UK for *Mr A J G England* of Padstow as a dual-purpose passenger and fishing vessel. In 1975 sold to *Mr M MacKenzie* of Ullapool, then to *Pentland Ferries, Wide Firth Ferry* in 1982, and *Orkney Islands Council* in 1986. Passenger only vessel. Generally operates feeder service between Pierowall (Westray) and Papa Westray.

GRAEMSAY Built by Ailsa Shipbuilding, Troon UK to operate between Stromness (Mainland), Moaness (Hoy) and Graemsay. Designed to offer an all year round service to these islands, primarily for passengers and cargo.

HOY HEAD Built by Appledore Shipbuilders, Appledore, UK to replace the THORSVOE on the 'South Isles' service (see above).

SHAPINSAY Built by Yorkshire Drydock, Hull, UK for the service from Kirkwall (Mainland) to Shapinsay.

THORSVOE Built by Campbeltown Shipyard, Campbeltown, UK for the 'South Isles' service (see above). In 1994 replaced by new HOY HEAD and became the main reserve vessel for the fleet.

VARAGEN Built by Cochrane Shipbuilders, Selby, UK for *Orkney Ferries*, a private company established to start a new route between Gills Bay (Caithness, Scotland) and Burwick (South Ronaldsay, Orkney). However, due to problems with the terminals it was not possible to maintain regular services. In 1991, the company was taken over by *OISC* and the VARAGEN became part of their fleet, sharing 'North Isles' services with the EARL SIGURD and the EARL THORFINN and replacing the freight vessel ISLANDER (494t, 1969).

PASSAGE EAST FERRY

THE COMPANY *Passage East Ferry Company Ltd* is an Irish Republic private sector company.

MANAGEMENT Managing Director Derek Donnelly, **Operations Manager** Conor Gilligan.

ADDRESS Barrack Street, Passage East, Co Waterford, Republic of Ireland.

TELEPHONE Administration +353 (0)51 382480, **Reservations** Not applicable.

FAX Administration +353 (0)51 382598, **Reservations** Not applicable.

INTERNET Email passageferry@eircom.net **Website** www.passageferry.com *(English)*

ROUTE OPERATED Passage East (County Waterford) – Ballyhack (County Wexford) (7 mins; *(1)*; frequent service).

1	EDMUND D	300t	68	9.0k	45.1m	143P	30C	-	BA	IR

EDMUND D Built by Philip Ltd, Dartmouth, UK as the SHANNON HEATHER for *Shannon Ferry Ltd* and used on their service between Killimer (County Clare) and Tarbert (County Kerry). Withdrawn from regular service in 1996 and, in 1997, sold to *Passage East Ferry* and renamed the EDMUND D. She entered service in January 1998.

PENTLAND FERRIES

THE COMPANY *Pentland Ferries* is a UK private sector company.

MANAGEMENT Managing Director Andrew Banks, **Marketing Manager** Linda Knott.

ADDRESS Pier Road, St Margaret's Hope, South Ronaldsay, Orkney KW172SW.

TELEPHONE Administration & Reservations +44 (0)1856 831226.

FAX Administration & Reservations +44 (0)1856 831614.

INTERNET Email sales@pentlandferries.co.uk **Website** www.pentlandferries.com *(English)*

ROUTE OPERATED Gills Bay (Scotland) – St Margaret's Hope (South Ronaldsay, Orkney) (1 hr; *(1,2)*; 3 per day).

Sound of Shuna *(Miles Cowsill)*

Cenwulf *(Andrew Cooke)*

C TOMS & SON LTD

THE COMPANY *C Toms & Son Ltd* is a British private sector company.

MANAGEMENT Managing Director Mr Allen Toms.

ADDRESS East Street, Polruan, Fowey, Cornwall PL23 1PB.

TELEPHONE Administration +44 (0)1726 870232.

FAX Administration +44 (0)1726 870318.

ROUTE OPERATED Fowey – Bodinnick (Cornwall) (5 mins; *(1,2)*; frequent).

1	GELLAN	50t	03	4.5k	36.0m	50P	10C	-	BA	UK
2	JENACK	60t	00	4.5k	36.0m	50P	15C	-	BA	UK

GELLAN, JENACK Built by *C Toms & Sons Ltd*, Fowey, UK. Self propelled and steered.

VALENTIA ISLAND FERRIES

THE COMPANY *Valentia Island Ferries Ltd* is an Irish Republic private sector company.

MANAGEMENT Manager Richard Foran.

ADDRESS Valentia Island, County Kerry, Republic of Ireland.

TELEPHONE Administration +353 (0)66 76141, **Reservations** Not applicable.

FAX Administration +353 (0)66 76377, **Reservations** Not applicable.

INTERNET Email reforan@indigo.ie **Website** indigo.ie/%7ecguiney/ferry.html *(English)*

ROUTE OPERATED Reenard (Co Kerry) – Knightstown (Valentia Island) (5 minutes; *(1)*; frequent service, 1st April – 30th September).

1	GOD MET ONS III	95t	63	-	43.0m	95P	18C	-	BA	IR

GOD MET ONS III Built by BV Scheepswerven Vh HH Bodewes, Millingen, Netherlands for *FMHE Res* of the Netherlands for a service across the River Maas between Cuijk and Middelaar. In 1987 a new bridge was opened and the service ceased. She was latterly used on contract work in the Elbe and then laid up. In 1996 acquired by *Valentia Island Ferries* and inaugurated a car ferry service to the island. Note: this island never had a car ferry service before. A bridge was opened at the south end of the island in 1970; before that a passenger/cargo service operated between Reenard Point and Knightstown.

WESTERN FERRIES

THE COMPANY *Western Ferries (Clyde) Ltd* is a British private sector company.

MANAGEMENT Managing Director Gordon Ross.

ADDRESSES Hunter's Quay, Dunoon PA23 8HJ.

TELEPHONE Administration +44 (0)1369 704452, **Reservations** Not applicable.

FAX Administration +44 (0)1369 706020, **Reservations** Not applicable.

INTERNET Email enquiries@western-ferries.co.uk **Website** www.western-ferries.co.uk *(English)*

ROUTE OPERATED McInroy's Point (Gourock) – Hunter's Quay (Dunoon) (20 mins; *(1,2,3,4)*; half hourly).

1●	SLEAT	466t	61	10.0k	39.9m	296P	30C	4/5L	BA	UK
2	SOUND OF SANDA	403t	64	10.0k	48.4m	220P	37C	4/5L	BA	UK
3	SOUND OF SCALPAY	403t	61	10.0k	48.4m	220P	37C	4/5L	BA	UK
4	SOUND OF SCARBA	489t	01	11.0k	49.5m	220P	40C	4/5L	BA	UK

5	SOUND OF SHUNA	489t	03	11.0k	49.5m	220P	40C	4/5L	BA	UK

SLEAT Built by NV Scheepswerf en Machinefabriek "de Merwede", Hardinxveld, Netherlands as the DE HOORN for the service between Maassluis and Rozenburg, across the 'Nieuwe Waterweg' (New Waterway) in The Netherlands. In 1988 she was purchased by *Western Ferries* and renamed the SOUND OF SLEAT. In 2003 laid up for sale. In 2004 renamed the SLEAT.

SOUND OF SANDA Built by Gutehoffnungshulte Sterkrade AG, Rheinwerft, Walsum, Germany as the GEMEENTEPONT 24 for *Amsterdam City Council* and operated from Centraal Station to the other side of the River IJ. In 1996 purchased by *Western Ferries* and renamed the SOUND OF SANDA.

SOUND OF SCALPAY Built by Arnhemsche Scheepsbouw Maatschappij NV, Arnhem, Netherlands as the Gemeentepont 23 for *Amsterdam City Council*. In 1995 sold to *Western Ferries* and renamed the SOUND OF SCALPAY.

SOUND OF SCARBA Built by Ferguson Shipbuilders, Port Glasgow, UK for *Western Ferries*.

SOUND OF SHUNA Built by Ferguson Shipbuilders, Port Glasgow, UK for *Western Ferries*.

WIGHTLINK

THE COMPANY *Wightlink* is a British private sector company, owned by the management. The routes and vessels were previously part of *Sealink (British Rail)* but were excluded from the purchase of most of the *Sealink* operations by *Stena Line AB* in 1990. They remained in *Sea Containers'* ownership until purchased by *CINVen* Ltd, a venture capital company in 1995. The company was the subject of a management buy-out financed by the *Royal Bank of Scotland* in 2001.

MANAGEMENT Chairman Michael Aiken, **Marketing Director** Janet Saville.

ADDRESS PO Box 59, Portsmouth PO1 2XB.

TELEPHONE Administration +44 (0)23 9281 2011, **Reservations** 08705 827744 (from UK only), +44 (0)23 9281 2011 (from overseas).

FAX Administration & Reservations +44 (0)23 9285 5257, **Telex** 86440 WIGHTLG.

INTERNET Email info@wightlink.co.uk **Website** www.wightlink.co.uk *(English)*

ROUTES OPERATED Conventional Ferries Lymington – Yarmouth (Isle of Wight) (approx 30 mins; *(1,2,3)*; half hourly), Portsmouth – Fishbourne (Isle of Wight) (approx 35 mins; *(8,9,10,11,12)*; half hourly or hourly depending on time of day). **Fast Passenger Ferries** Portsmouth – Ryde (Isle of Wight) (passenger only) (approx 15 mins; *(4,5,6,7)*; half hourly/hourly).

1	CAEDMON	764t	73	9.5k	57.9m	512P	58C	6L	BA	UK
2	CENRED	761t	73	9.5k	57.9m	512P	58C	6L	BA	UK
3	CENWULF	761t	73	9.5k	57.9m	512P	58C	6L	BA	UK
4»p	FASTCAT RYDE	478t	96	34.0k	40.0m	361P	0C	0L	-	UK
5»p	FASTCAT SHANKLIN	482t	96	34.0k	40.0m	361P	0C	0L	-	UK
6»p	OUR LADY PAMELA	312t	86	28.5k	29.5m	395P	0C	0L	-	UK
7»p	OUR LADY PATRICIA	312t	85	28.5k	29.5m	395P	0C	0L	-	UK
8	ST CATHERINE	2038t	83	12.5k	77.0m	771P	142C	12L	BA	UK
9	ST CECILIA	2968t	86	12.5k	77.0m	771P	142C	12L	BA	UK
10	ST CLARE	5359t	01	13.0k	86.0m	878P	186C	-	BA	UK
11	ST FAITH	3009t	89	12.5k	77.0m	771P	142C	12L	BA	UK
12	ST HELEN	2983t	83	12.5k	77.0m	771P	142C	12L	BA	UK

CAEDMON Built by Robb Caledon Shipbuilders Ltd, Dundee, UK for *Sealink* for the Portsmouth – Fishbourne service. In 1983 transferred to the Lymington – Yarmouth service.

CENRED, CENWULF Built by Robb Caledon Shipbuilders Ltd, Dundee, UK for *Sealink* for the Lymington – Yarmouth service.

FASTCAT RYDE Kværner Fjellstrand Flyingcat 40m built at Singapore as the WATER JET 1 for

Waterjet Netherlands Antilles and operated in the Philippines. In 1999 renamed the SUPERCAT 17. In summer 2000 sold to *Wightlink* and renamed the FASTCAT RYDE. After modifications, entered service on the Portsmouth – Ryde route in autumn 2000.

FASTCAT SHANKLIN Kværner Fjellstrand Flyingcat 40m built at Singapore as the WATER JET 2 for *Waterjet Netherlands Antilles* and operated in the Philippines. In 1999 renamed the SUPERCAT 18. In summer 2000 sold to *Wightlink* and renamed the FASTCAT SHANKLIN. After modifications, entered service on the Portsmouth – Ryde route in autumn 2000.

OUR LADY PAMELA, OUR LADY PATRICIA InCat 30 m catamarans built at Hobart, Tasmania, Australia for *Sealink* for the Portsmouth – Ryde service.

ST CATHERINE, ST HELEN Built by Henry Robb Caledon Yard, Leith, UK for *Sealink* for the Portsmouth – Fishbourne service.

ST CECILIA, ST FAITH Built by Cochrane Yard, Selby, UK for *Sealink* for the Portsmouth – Fishbourne service.

ST CLARE Built by Stocznia Remontowa, Gdansk, Poland for *Wightlink* for the Portsmouth – Fishbourne service. She is a double ended ferry with a central bridge.

WOOLWICH FREE FERRY

THE COMPANY The *Woolwich Free Ferry* is operated by the *London Borough of Greenwich*, a British municipal authority on behalf of *Transport for London*.

MANAGEMENT Ferry Manager Capt P Deeks.

ADDRESS New Ferry Approach, Woolwich, London SE18 6DX.

TELEPHONE Administration +44 (0)20 8921 5786, +44 (0)20 8921 5967, **Reservations** Not applicable.

FAX Administration +44 (0)20 8316 6096, **Reservations** Not applicable.

INTERNET Email peter.deeks@greenwich.gov.uk

Website www.greenwich.gov.uk/Greenwich/Travel/LocalTravelServices/WoolwichFerry.htm *(English)*

ROUTE OPERATED Woolwich – North Woolwich (free ferry) (5 mins; *(1,2,3)*; every 9 mins (weekdays – two ferries in operation), every 16 mins (weekends – one ferry in operation)). Note: one ferry is always in reserve/under maintenance.

1	ERNEST BEVIN	738t	63	8.0k	56.7m	310P	32C	6L	BA	UK
2	JAMES NEWMAN	738t	63	8.0k	56.7m	310P	32C	6L	BA	UK
3	JOHN BURNS	738t	63	8.0k	56.7m	310P	32C	6L	BA	UK

ERNEST BEVIN, JAMES NEWMAN, JOHN BURNS Built by Robb Caledon Shipbuilders Ltd, Dundee, UK for the *London County Council* who operated the service in 1963. In 1965 ownership was transferred to the *Greater London Council*. Following the abolition of the *GLC* in April 1986, ownership was transferred to the *Department of Transport* and in 2001 to *Transport for London*. The *London Borough of Greenwich* operate the service on their behalf. An alternative loading is 6m x 18m articulated lorries cars and 14 cars; lorries of this length are too high for the nearby northbound Blackwall Tunnel.

seafrance.com
SEAFRAN

SeaFrance Nord Pas-de-Calais *(Miles Cowsill)*

section 3

freight-only ferries

BRITTANY FERRIES

THE COMPANY *See section 1.* The freight division of *Brittany Ferries* traded as *Truckline Ferries* until 2002.

MANAGEMENT Managing Director David Longden, **Freight Director** John Clarke.

ADDRESS New Harbour Road, POOLE, Dorset BH15 4AJ.

TELEPHONE Administration +44 (0)8709 013300, **Reservations** +44 (0)8709 040200.

FAX Administration & Reservations +44 (0)1202 679828, **Telex** 41744, 41745.

INTERNET Website www.brittanyferriesfreight.co.uk *(English)*

ROUTES OPERATED Cherbourg (***Winter*** *dep: 09.30 Wed, Fri, Sun, 18.30 Tue, Thu, Sat, 23.45 Mon, Wed, Fri, Sun,* ***Summer*** *02.00 Mon, Sat, Sun, 09.30 Tue, Thu, 14.30 Fri, Sat, Sun, 18.30 Mon, Wed, 23.45 Tue, Thu)* – Poole *(**Winter** dep: 16.00 Mon, Wed, Fri, Sun, 08.30, 23.45 Tue, Thu, Sat,* ***Summer*** *dep: 16.00 Tue, Thu, 08.30, 23.45 Mon, Wed, 07.30, 23.45 Fri, Sat, Sun)* (4 hrs 30 mins; *(1)*; 1/2 per day). **Note** Operates with the passenger vessel BARFLEUR to provide three or four sailings every 24 hrs.

1	COUTANCES	6507t	78	17.0k	125.2m	58P	-	58T	BA	FR

COUTANCES Built by Ateliers et Chantiers du Havre, Le Havre, France for *Truckline Ferries* for their Cherbourg – Poole service. In 1986 lengthened to increase vehicle capacity by 34%.

CALEDONIAN MACBRAYNE

THE COMPANY, MANAGEMENT, ADDRESS, TELEPHONE & INTERNET See section 2.

ROUTE OPERATED Ullapool – Stornoway (Lewis) (3 hours; *(1)*; 1 per day) (no fixed schedule – run according to demand).

1	MUIRNEAG	5801t	79	15.5k	105.6m	0P	-	54T	AS	UK

MUIRNEAG Built by Frederikshavn Værft A/S, Frederikshavn, Denmark as the MERCANDIAN CARRIER II for *Mercandia* of Denmark and used on a variety of services. In 1983 she was briefly renamed ALIANZA and between 1984 and 1985 she carried the name CARRIER II. In 1985 sold to *P&O*, renamed the BELARD and used on *Northern Ireland Trailers* services between Ardrossan and Belfast (later Larne), subsequently becoming part of *Pandoro*. In 1993 she was chartered to *IOMSP* subsidiary *Mannin Line* to inaugurate a new service between Great Yarmouth and IJmuiden. In 1994 she was purchased by *IOMSP*; however, in 1995 the *Mannin Line* service ceased and she was chartered back to *Pandoro*. At the end of 1995 she was returned to *IOMSP*. In 1996, after deputising for the PEVERIL, she was briefly chartered to *Exxtor Ferries* (operating between Immingham and Rotterdam) and then laid up. In 1997, she again deputised for the PEVERIL, followed by a short period of charter to *P&O European Ferries*. In 1997 she returned to *IOMSP* and replaced the PEVERIL as the main freight vessel. In 1998 she was sold to the *Aabrenaa Rederi* and operated between Åbenrå and Klaìpeda. This service ended in 199 and she was used on a number of short term charters. In spring 2002 chartered to *Ferryways* and operated between Ipswich and Ostend. In autumn 2002 she was chartered to *Caledonian MacBrayne*, renamed the MUIRNEAG and placed on the Ullapool – Stornoway service, replacing the HASCOSAY (see *NorthLink*).

Coutances *(Brian Mitchell)*

Muirneag *(CalMac)*

CHANNEL FREIGHT FERRIES

THE COMPANY *Channel Freight Ferries Ltd* is a UK private sector company, wholly owned by *Clarkson plc.*

MANAGEMENT Managing Director Simon Taylor, **General Manager – UK** Kevin Miller.

ADDRESS *UK* 30 Berth, European Way, Eastern Docks, Southampton SO14 3XD, ***France*** Quai de Radicatel, Saint-Jean-de-Folleville, 76170 Lillebonne.

TELEPHONE Administration & Reservations *UK* +44 (0)2380 205 900, ***France*** +33 2 35 39 01 01.

FAX Administration & Reservations *UK* +44 (0)2380 205 901, ***France*** +33 2 35 39 39 70

INTERNET Email info@channelfreight.com **Website** www.channelfreight.com *(English, French)*

ROUTE OPERATED Southampton *(dep: 21.00 Sun & Fri, 07.45 Tue, 08.15 Wed, 08.45 Thu)* – Radicatel (Port of Rouen) *(dep: 20.30 Mon, 21.00 Tue, 21.30 Wed, 22.00 Thu & Sat)* (9 hrs; *(1)*; 1 per day). **Note** schedule and ship liable to change.

1	CFF SEINE	9071t	84	15.0k	121.5m	12P	-	90T	A	RO

CFF SEINE Built by Santierul Naval SA, Galatz, Romania as the BALDER FJORD for *K/S A/S Balder RO/RO No 2* of Norway. In 1986 acquired by *Navrom* of Romania and renamed the BAZIAS 1. In 1990 transferred to *Romline* of Romania and subsequently sold to *Octogon Shipping* of Romania. In 1996 chartered to *Ignazio Messina* of Italy and later renamed the JOLLY ARANCIONE. In late 1997 chartered to *Dart Line* and renamed the DART 1. In late 1999 charter ended, and she was briefly chartered to *Merchant Ferries*, although she was taken back on short term charter by *Dart Line* in February 2000. In 2001 renamed the BAZIAS 1 and in April chartered to *Cobelfret Ferries* to operate between Rotterdam and Purfleet. In autumn she transferred to the Zeebrugge -Purfleet service. In March 2002 chartered to *Dart Line* to operate between Dartford and Dunkerque (not inter-worked with the Vlissingen service). In summer 2002 charter ended. Re-chartered in autumn 2002 and worked additional sailings from Dartford to Zeebrugge. From November 2002 she operated a triangular Dartford – Dunkerque – Zeebrugge – Dartford service. From January 2003 resumed service between Dartford and Zeebrugge only. Later operated mainly between Dartford and Dunkerque. In 2004 chartered to *Channel Freight Ferries* and renamed the CFF SEINE. Later in 2004 purchased by *Channel Freight Ferries.*

COBELFRET FERRIES

THE COMPANY *Cobelfret Ferries nv* is a Belgian private sector company, a subsidiary of *Cobelfret nv* of Antwerp.

MANAGEMENT Operations Manager (Belgium) Marc Vandersteen, **UK *Purfleet and Dagenham services*** Cobelfret Ferries UK Ltd – **General Manager, Line & Agency Division** Nick Kavanagh, ***Immingham Services*** Cobelfret Ferries UK Ltd (Immingham Branch) – **General Manager** Peter Kirman.

ADDRESS *Belgium* 8380 Zeebrugge, Belgium, ***UK Purfleet*** Purfleet Thames Terminal, London Road, Purfleet, Essex RM19 1RP, ***UK Immingham (Killingholme)*** Cobelfret Ferries UK (Killingholme Brach), Humber Sea Terminal, Clough Lane, North Lincolnshire DN40 3JS.

TELEPHONE Administration & Reservations *Belgium* +32 (0)50 502243, ***UK (Purfleet)*** +44 (0)1708 891199, ***(Immingham (Killingholme))*** +44 (0)1469 542500.

FAX Administration & Reservations *Belgium* +32 (0)50 502219, ***UK (Purfleet)*** +44 (0)1708 890853, ***(Immingham (Killingholme))*** +44 (0)1469 542501.

INTERNET Email *Zeebrugge* pur.cobzee@cobelfretferries.be

Purfleet nick.kavanagh@cobelfretfrerries.co.uk **Website:** www.cobelfret.com *(English)*

ROUTES OPERATED Zeebrugge *(dep: 04.00 Tue-Fri, 10.00 Mon-Fri, 16.00 Mon-Fri, 22.00 Mon-Fri)*

SECTION 3 – FREIGHT ONLY FERRIES

– Purfleet *(dep: 06.00 Tue-Fri, 12.00 Mon-Fri, 18.00 Mon-Fri, 23.00 Mon-Fri (Note: weekend service run subject to demand and are liable to vary))* (9 hrs; (*3,5,12,19,20*); 4 per day), Vlissingen *(dep: 02.00 Tue-Fri, 10.00 Mon-Fri, 18.00 Mon-Fri)* – Dagenham *(dep: 01.00 Tue-Fri, 09.00 Mon-Fri, 17.00 Mon-Fri)* (contract service for Ford Motor Company) (11 hrs; (*8,9,17,18*); 3 per day), Rotterdam (Brittanniehaven) *(dep: 17.00 Sat, 20.00 Mon-Fri)* – Purfleet *(dep: 16.00 Sat, 19.00 Mon-Fri)* (14 hrs; (*1,11*); 1 per day), Zeebrugge *(dep: 17.00 Sat, 19.00 Mon-Fri)* – Immingham (Killingholme) *(17.00 Sat, 19.00 Mon-Fri)* (14 hrs; (*5,7*) (Sunday sailings as required)); 1/2 per day), Rotterdam (Brittanniehaven) *(dep: 17.30)* – Immingham (Killingholme) *(dep: 17.30)* (14 hrs; (*4,13*); 1 per day), Immingham (Killingholme) *(dep: Tue 01.00, Thu 10.00, Fri 11.00, Sun 22.00)* – Gothenburg *(dep: Tue 17.00, Wed 19.00, Sat 05.00, Sun 06.00)* (36 hrs; (*3,10*); 4 per week), Zeebrugge *(dep: 18.00 Fri)* – Esbjerg *(dep: 00.01 Sun)* (24 hrs; (*varies*), Zeebrugge *(dep: 06.00 Sun, 09.00 Mon, 12.00 Tue, 18.00 Wed, 18.00 Thu, 22.00 Fri)* – Gothenburg *(dep: 16.00 Sun, 00.01 Tue, 00.01 Wed, 04.00 Thu, 13.00 Fri, 13.00 Sat)* (33-38 hrs; *(14,15,16)*; 6 per week) (this service is operated by *Wagenborg* of the Netherlands for the *Stora-Enso* paper and board group, for the conveyance of their products. *Cobelfret Ferries* act as handling agents at Zeebrugge and market the surplus capacity on the vessels, which is available for general ro-ro traffic. Although this route is strictly outside the scope of this book it is included for the sake of completeness). Correct at 30th April 2005; ships are quite frequently moved between routes so the above may have changed by the time this book is published.

1	AMANDA	14715t	78	14.5k	172.9m	12P	-	133T	A	BZ
2	ANNA ODEN	16947t	78	15.0k	170.3m	12P	180C	160T	A	SW
3	BRITTA ODEN	16950t	79	15.0k	170.3m	12P	180C	160T	A	SW
4	CATHERINE	21287t	02	20.0k	182.2m	12P		200T	A2	LX
5	CELANDINE	23986t	00	18.0k	162.5m	12P	630C	157T	A	BS
6	CELESTINE	23986t	96	17.8k	162.5m	24P	630C	157T	A	BS
7	CLEMENTINE	23986t	97	17.8k	162.5m	24P	630C	157T	A	LX
8	CYMBELINE	11866t	92	14.5k	147.4m	8P	350C	100T	A2	LX
9	EGLANTINE	10035t	89	14.5k	147.4m	8P	350C	100T	A2	LX
10	EVA ODEN	16950t	79	15.0k	170.3m	12P	180C	160T	A	SW
11	LOUISE RUSS	18400t	00	23.5k	174.0m	12P	-	171T	A	GI
12	MARABOU	10931t	78	17.0k	161.4m	12P	-	102T	A2	BZ
13	MELUSINE	23987t	99	18.0k	162.5m	12P	630C	157T	A	LX
14	SCHIEBORG	21005t	00	17.0k	183.4m	12P	-	180T	A	NL
15	SLINGEBORG	21005t	00	17.0k	183.4m	12P	-	180T	A	NL
16	SPAARNEBORG	21005t	00	17.0k	183.4m	12P	-	180T	A	NL
17	SYMPHORINE	10030t	88	14.5k	147.4m	8P	350C	100T	A2	LX
18	UNDINE	11854t	91	14.5k	147.4m	8P	350C	100T	A2	LX
19	VALENTINE	23987t	99	18.0k	162.5m	12P	630C	157T	A	BS
20	VICTORINE	23987t	00	18.0k	162.5m	12P	630C	157T	A	BS

AMANDA Built by Paul Lindenau Werft, Kiel, Germany as the MERZARIO PERSIA for *Merzario Line* of Italy and used on services between Italy and the Middle East. In 1986 she was chartered to *Grimaldi* of Italy and renamed the PERSIA, continuing on Middle East services. In 1988 she was sold to *Eimskip* of Iceland and renamed the BRUARFOSS. She was used on their service between Reykjavik, Immingham, Hamburg and Rotterdam. In 1996, the ro-ro service was replaced by a container only service and she was withdrawn. She was renamed the VEGA and was placed a number of short term charters including *Suardiaz* of Spain and *Fred. Olsen Lines*. In 1998, she was sold to *Cobelfret* and renamed the AMANDINE. Used mainly on the Rotterdam – Immingham service until 2002 when she was transferred to the Rotterdam – Purfleet route. In 2003 sold and renamed the AMANDA. Chartered back.

ANNA ODEN Built by Öresundvarvet Ab, Landskrona, Sweden as the ANNA ODEN for *AB Norsjöfrakt* of Sweden and chartered to *Oden Line* of Sweden for North Sea services, in particular associated with the export of Volvo cars and trucks from Göteborg. In 1980 *Oden Line* was taken over by *Tor Lloyd AB*, a joint venture between *Tor Line* and *Broströms AB* and the charter transferred to them.

moving to *Tor Line* in 1981 when *DFDS* took over. In 1987 she was lengthened and on re-entry into service in early 1988 was renamed the TOR FLANDRIA and became regular vessel on the Göteborg – Ghent (Belgium) service, largely operated for Volvo. In 1999 the charter was ended and she was renamed the SOUTHERN CARRIER. She was chartered to a number of operators including *Cobelfret Ferries* and *Flota Suardiaz*. In 2002 she was renamed the ANNA ODEN. In 2003 she was sold to *Cobelfret Ferries*. In 2004 inaugurated a new Immingham – Gothenburg service. In 2005 moved to the Zeebrugge – Purfleet route.

BRITTA ODEN Built by Öresundvarvet Ab, Landskrona, Sweden as the BRITTA ODEN for *AB Norsjöfrakt* (later *Bylock & Norsjöfrakt*) of Sweden and chartered to *Oden Line* of Sweden for North Sea services, in particular associated with the export of Volvo cars and trucks from Göteborg. In 1980 *Oden Line* was taken over by *Tor Lloyd AB*, a joint venture between *Tor Line* and *Broströms AB* and the charter transferred to them, moving to *Tor Line* in 1981 when *DFDS* took over. In 1987 she was enlarged and on re-entry into service in early 1988 was renamed the TOR SCANDIA and became regular vessel on the Göteborg – Gent (Belgium) service. In 1998 renamed the BRITTA ODEN. In 1999 the charter was ended and she was chartered to *Flota Suardiaz*. In September 2003 she was chartered to *Cobelfret Ferries* and placed on the Rotterdam – Purfleet service. In October 2003 purchased by *Cobelfret Ferries*. In 2004 inaugurated a new Immingham – Gothenburg service.

CATHERINE Built as the ROMIRA by Zhonghua Shipyard, Zhonghua, China for *Dag Engström Rederi* of Sweden. For six months engaged on a number of short term charters, including *Cobelfret Ferries* who used her on both the Rotterdam – Immingham and Zeebrugge – Purfleet routes. In September 2002 purchased by *Cobelfret Ferries* and in November 2002, renamed the CATHERINE and started on the Rotterdam – Immingham service. In spring 2003 chartered to the *US Defense Department* to convey materials to the Persian Gulf. Returned in late summer.

CELANDINE, VALENTINE, VICTORINE Built by Kawasaki Heavy Industries, Sakaide, Japan for *Cobelfret*. Similar to the CLEMENTINE. The CELANDINE was originally to be called the CATHERINE and the VICTORINE the CELANDINE. The names were changed before delivery. They are generally used on the Zeebrugge – Purfleet service.

CELESTINE Built by Kawasaki Heavy Industries, Sakaide, Japan as the CELESTINE. In 1996 chartered to the *British MoD* and renamed the SEA CRUSADER. She was originally expected to return to *Cobelfret Ferries* in early 2003 and resume the name CELESTINE; however the charter was extended because of the Iraq crisis. Returned in September 2003 and placed on the Zeebrugge – Immingham service.

CLEMENTINE Built by Kawasaki Heavy Industries, Sakaide, Japan for *Cobelfret*. Currently used on the Zeebrugge – Immingham service.

CYMBELINE, EGLANTINE, SYMPHORINE, UNDINE Built by Dalian Shipyard, Dalian, China for *Cobelfret*. Used on the Dagenham – Vlissingen route.

EVA ODEN Built by Öresundvarvet AB, Landskrona, Sweden as the EVA ODEN for *AB Norsjöfrakt* (later *Bylock & Norsjöfrakt*) of Sweden and chartered to *Oden Line* of Sweden for North Sea services, in particular associated with the export of Volvo cars and trucks from Gothenburg. In 1980 *Oden Line* was taken over by *Tor Lloyd AB*, a joint venture between *Tor Line* and *Broströms AB* and the charter transferred to them, moving to *Tor Line* in 1981 when *DFDS* took over. In 1987 she was enlarged and on re-entry into service in early 1988 was renamed the TOR BELGIA and became regular vessel on the Gothenburg – Ghent (Belgium) service. In 1998 renamed the EVA ODEN and in 1999 the charter was terminated. In 2000 she was chartered to *Cobelfret Ferries*. Until 2003, generally used on the Zeebrugge – Purfleet service. In 2003 transferred to the Rotterdam – Immingham service, replacing the CATHERINE. In July 2003 transferred to the Rotterdam – Purfleet route. In September replaced by the BRITTA ODEN and, on return from refit placed on the Zeebrugge – Purfleet route. In 2005 transferred to the Immingham – Gothenburg route.

LOUISE RUSS Launched by J J Sietas Schiffswerft, Hamburg, Germany as the LOUISE RUSS for *Ernst Russ* of Germany. On completion, renamed the PORTO EXPRESS and chartered to *ROROExpress* to operate between Southampton, Oporto and Tangier. The service ceased in autumn 2001 and she was returned to her owners and resumed the name LOUISE RUSS. In 2002 chartered to *Cobelfret Ferries* and placed on the Rotterdam – Immingham service, replacing the AMANDINE. In 2004 transferred

Commodore Goodwill *(Miles Cowsill)*

Victorine *(Nick Widdows)*

to the Rotterdam – Purfleet service. In January 2005 temporarily transferred to the Zeebrugge – Gothenburg route to replace the fire damaged SCHIEBORG. Expected to return to Purfleet in June 2005.

MARABOU Built by AB Lödöse Varv, Lödöse, Sweden as the VALLMO for the *Johansson Group* of Sweden and undertook a variety of charters. In 1982 she was sold to *Cobelfret* and renamed the MATINA. In 1984 renamed the LOVERVAL. In recent years has been chartered out for periods. In 2003 sold and renamed the MARABOU; chartered back. Usually used on the Zeebrugge – Purfleet service but at time of going to press on charter to *Van-Uden RoRo.*

MELUSINE Built by Kawasaki Heavy Industries, Sakaide, Japan for *Cobelfret.* Similar to the CLEMENTINE. Currently being used on the Rotterdam – Immingham service.

SCHIEBORG, SLINGEBORG, SPAARNEBORG Built by Flender Werft AG, Lübeck, Germany for *Wagenborg* of The Netherlands and time chartered to *Stora-Enso* to operate between Zeebrugge and Gothenburg. The SCHIEBORG was badly damaged by fire in heavy seas in January 2005 and is still under repair at time of going to press.

Under Construction

21	PAULINE	50000t	06	21.7k	200.0m	12P	656C	258T	A	LX
22	YASMINE	50000t	07	21.7k	200.0m	12P	656C	258T	A	LX

PAULINE, YASMINE Under construction by Flensburger Schiffbau-Gesellschaft, Flensburg, Germany to operate on the Rotterdam – Immingham or Zeebrugge – Immingham routes.

CONDOR FERRIES

THE COMPANY *Condor Ferries Ltd* is a Channel Islands private sector company owned by the *Condor Group,* Guernsey which is owned by the *Royal Bank of Scotland.*

MANAGEMENT Managing Director Robert Provan, **Freight Director** Jeff Vidamour.

ADDRESS Head Office PO Box 10, New Jetty Offices, White Rock, St Peter Port, Guernsey GY1 3AF, Sales and Marketing Condor House, New Harbour Road South, Hamworthy, POOLE BH15 4AJ.

TELEPHONE Administration & Reservations +44 (0)1481 728620.

FAX Administration & Reservations +44 (0)1481 728521.

INTERNET Email jeff.vidamour@condorferries.co.uk **Websites:** www.condorferries.co.uk *(English)*

www.condorferries.fr *(French, English, Dutch, German)*

ROUTE OPERATED Portsmouth *(dep: 09.00*, 19.30)* – Guernsey *(dep: 04.00, 17.30*)* (6 hrs 30 min) – Jersey *(dep: 08.15, 21.30*)* (10 hrs 30 min; *(1)*; 2 per day) (*operated by ro-pax ferry COMMODORE CLIPPER – see Section 1), Guernsey *(dep: 07.00 Sat)* – Jersey *(dep: 11.00 Sat)* – St Malo *(arr: 14.00 Sat, dep: 17.00 Sat)* – Jersey *(arr: 06.00 Sun)* – Guernsey *(arr: 03.00 Mon)*; *(1)*; 1 per week).

1	COMMODORE GOODWILL	11166t	96	18.3k	126.4m	12P	-	92T	A	BS

COMMODORE GOODWILL Built by Koninklijke Scheldegroep BV, Vlissingen, Netherlands for *Commodore Ferries.*

DART LINE

THE COMPANY *Dart Line Ltd* is a British private sector company, now owned by *The Bidvest Group Ltd* of South Africa It took over the Dartford – Vlissingen service from *Sally Ferries* in 1996.

MANAGEMENT Managing Director Ron Herman, **Head of Dart Line** Jean Claude Raes, **Sales Director** Nick Pank.

ADDRESS Dart Terminal, Crossways Business Park, Dartford, Kent DA2 6QB.

TELEPHONE Administration & Reservations +44 (0)1322 281122.

FAX Administration & Reservations +44 (0)1322 281133.

INTERNET Email sales@dartline.co.uk **Website** www.dartline.co.uk *(English)*

ROUTES OPERATED Dartford *(dep: 08.30 Tue-Fri, 20.30 daily)* – Zeebrugge *(dep: 09.30 Tue-Fri, 20.30 Daily)* (8 hrs 30 mins; *(3,4)*; up to 1/2 per day), Dartford *(dep: 09.00 Tue-Fri, 20.00 Daily)* – Vlissingen *(dep: 09.30 Tue-Fri, 21.30 Daily)* (9 hrs 30 mins; *(1,2)*; 1/2 per day).

1	DART 3	9080t	84	15.0k	121.5m	12P	-	90T	A2	BD
2	DART 4	9080t	84	15.0k	121.5m	12P	-	90T	A2	BD
3	DART 8	22748t	80	18.0k	178.5m	12P	-	155T	A	BD
4	DART 9	22748t	80	18.0k	178.5m	12P	-	155T	A	BD

DART 3 Built by Santierul Naval, Galatz, Romania as the BALDER STEN for *K/S A/S Balder RO/RO No 2* of Norway (part of the *Parley Augustsson* group). In 1995 acquired by *Navrom* of Romania and renamed the BAZIAS 3. In 1991 chartered to *Sally Ferries* for the Ramsgate – Ostend freight service and subsequently purchased by a joint *Sally Ferries/Romline* company. In 1993 renamed the SALLY EUROROUTE and re-registered in The Bahamas. In October 1996 she was chartered to *Belfast Freight Ferries* and renamed the MERLE. In 1997 *Sally Ferries'* interests in her were purchased by *Jacobs Holdings*. In January 2000 she joined *Dart Line* and was placed on the Vlissingen service, being renamed the DART 3. In autumn 2000 she was chartered to *NorseMerchant Ferries* and placed again on the Heysham – Belfast service. In autumn 2001 returned to *Dart Line*. Currently operates on the Dartford – Vlissingen route.

DART 4 Built by Santierul Naval, Galatz, Romania as the BALDER BRE for *K/S A/S Balder RO/RO No 2* of Norway. Later in 1985 acquired by *Navrom* of Romania and renamed the BAZIAS 4. In 1991 chartered to *Sally Ferries* for the Ramsgate – Ostend freight service and subsequently purchased by *Rosal SA*, a joint *Sally Ferries/Romline* company. In 1993 renamed the SALLY EUROLINK and re-registered in The Bahamas. In 1997 *Sally Ferries*' interests in her were purchased by *Jacobs Holdings*. She was later transferred to *Dart Line* and renamed the DART 4. In 1998 she was chartered to *Belfast Freight Ferries*. She returned to *Dart Line* in February 1999 and operates on the Dartford – Vlissingen route.

DART 8 Built by Kawasaki Heavy Industries, Sakaide, Japan as the XI FENG KOU, a deep sea ro-ro/container ship for *China Ocean Shipping Company* of the People's Republic of China for service between the USA, Australia and New Zealand. In 1999, purchased by *Jacobs Holdings*. After delivery, she was converted in Nantong, China to short sea ro-ro specification, including the fitting of a stern ramp (replacing the quarter ramp) and luxury accommodation for 12 drivers and entered service in August 1999 on the Dartford – Zeebrugge service. In 2003 chartered to the *British MoD* for service in the Persian Gulf. Returned in September 2003 and resumed service on the Dartford – Zeebrugge route.

DART 9 Built by Kawasaki Heavy Industries, Sakaide, Japan as the GU BEI KOU. Sister vessel of the DART 8. She entered service in September 1999 on the Dartford – Zeebrugge service. In 2003 chartered to the *British MoD* for service in the Persian Gulf. Returned in September 2003 and resumed service on the Dartford – Zeebrugge route.

DFDS TOR LINE

THE COMPANY *DFDS Tor Line* is primarily a ro-ro operator on the North Sea and Baltic Sea. It is part of *DFDS A/S* which was formed in 1866 and is today quoted on the Copenhagen Stock Exchange. The *DFDS A/S* group consists of companies in Denmark, Sweden, Norway, the United Kingdom, the Netherlands, Belgium, Germany and Lithuania.

MANAGEMENT Managing Director Björn Petrusson, **Managing Director DFDS Tor Line plc** Jens Skibsted Nielsen.

ADDRESS ***Denmark (Head Office)*** Sundkrogsgade 11, DK-2100 Copenhagen Ø, ***UK*** Nordic House, Western Access Road, Immingham Dock, Immingham, South Humberside DN40 2LZ.

TELEPHONE Administration & Reservations ***Denmark (Head Office)*** +45 33 42 33 00, ***UK*** +44 (0)1469 575231.

FAX Administration & Reservations ***Denmark*** +45 33 42 33 01, ***UK*** +44 (0)1469 552690.

INTERNET Email info@dfdstorline.com **Website** www.dfdstorline.com *(English)*

ROUTES OPERATED

Esbjerg *(dep: 21.30 Mon-Sat)* – Immingham *(dep: 20.30 Mon-Sat)* (18 hrs)); *(7,23)*; 6 per week, Cuxhaven *(dep: 19.00 Mon, 19.00 Tue, 02.00 Thu, 19.00 Fri, 09.00 Sat)* – Immingham *(dep: 11.00 Sun, 22.30 Tue, 22.30 Wed, 04.30 Fri, 22.30 Sat)* (22 hrs; *(9,16)*; 9 per week), Gothenburg *(dep: 19.00 Tue, 20.00 Fri)* – Harwich *(dep: 04.00 Sun, Thu)*; 30 hrs; *(22)*; 2 per week), Gothenburg *(dep: 21.00 Sun-Fri, 18.00 Sat)* – Immingham *(dep: 04.00 Sun-Fri, 10.00 Sat)* – (26 hrs); *(5,11,19)*; 7 per week), Rotterdam (Maasvlakte) *(dep: 19.00 Mon, 18.00 Tue-Fri, 17.00 Sat)* – Immingham *(dep: 18.15 Mon-Fri, 17.00 Sat)* (14 hrs 30 mins); *(14,17)*; 6 per week), Zeebrugge *(dep: 17.30 Mon-Sat)* – Immingham *(dep: 17.00 Mon-Sat)* (15 hrs; *(1,8)*; 6 per week), Gothenburg *(dep: 01.00 Tue-Sat, 21.00 Sat)* – Brevik (Norway) *(dep: 11.00 Mon, 11.00 Sat,)* – Ghent (Belgium) *(dep: 03.00 Tue-Sat (Brevik served by Thu* & Sat departures), 23.00 Sat* (Gothenburg 37 hrs, Brevik 33 hrs; *(10,15,20,21)*; 6 per week) (*calls at Gothenburg before Brevik), Brevik *(dep: 03.00 Mon, 20.00 Thu)* – Kristiansand *(dep: 11.00 Mon, 02.30 Thu)* – Immingham *(dep: 20.30 Tue, 14.00 Sat)* (approx 27-29 hours Norwegian Port – Immingham; *(15)*; 2 per week) (**Note** Tue ex Immingham operates Immingham – Kristiansand – Brevik – Immingham, Sat ex Immingham operates Immingham – Brevik – Kristiansand – Immingham), Fredericia *(dep: 22.00 Tue, 02.30 Sat)* – Copenhagen *(dep: Wed 09.00, Sat 15.00)* – Klaìpeda (Lithuania) *(dep: 21.00 Sun, 13.00 Thu)*, *(4)*; 2 per week), *DFDS Tor Line* also operates a Ro-pax service between Lübeck and Riga. See Section 6.

Space is also used for freight on *DFDS Seaways/DFDS Tor Line* passenger vessels between Harwich and Esbjerg, Cuxhaven and Harwich (3 per week or alternate days during the summer), Gothenburg and Kristiansand/Newcastle (2-3 per week) and IJmuiden – Newcastle (daily). An additional freight service was operated between Esbjerg and Harwich during summer 2004 and this may be repeated in summer 2005.

Note Non-UK routes shown above are strictly outside the scope of this book but are shown for the sale of completeness.

1	AURORA	20391t	82	18.5k	155.0m	12P	-	160T	A	NO
2	TANGO	9983t	84	15.5k	138.5m	12P	-	69T	A	RO
3	TOR ANGLIA	17492t	77	16.0k	171.9m	12P	-	180T	A	DK
4	TOR BALTICA	14374t	78	18.5k	163.6m	12P	-	184T	A	BZ
5	TOR BEGONIA	32289t	04	22.5k	199.8m	12P	-	280T	AS	SW
6	TOR BELGIA	21491t	78	18.5k	193.3m	12P	200C	194T	AS	SW
7	TOR BRITANNIA	24196t	00	21.1k	197.5m	12P	-	200T	A	DK
8	TOR CIMBRIA	12189t	86	17.0k	145.0m	12P	-	142T	A	UK
9	TOR DANIA	21491t	78	19.5k	193.6m	12P	200C	192T	AS	SW
10	TOR FLANDRIA	33652t	81	19.7k	193.6m	12P	300C	204T	A	SW
11	TOR FREESIA	32289t	05	22.5k	199.8m	12P	-	280T	AS	SW

Tor Minerva *(Cees de Bijl)*

Tor Primula *(Cees de Bijl)*

12	TOR FUTURA	18725t	96	19.7k	183.3m	12P	-	164T	AS	DK
13	TOR HOLLANDIA	12259t	73	16.0k	163.5m	12P	-	110T	A	NO
14	TOR HUMBRIA	20165t	78	17.5k	177.4m	12P	-	154T	A	NO
15	TOR MAGNOLIA	32289t	03	22.5k	199.8m	12P	-	280T	AS	DK
16	TOR MAXIMA	17068t	78	17.0k	176.2m	12P	-	194T	A	NO
17	TOR MINERVA	21215t	78	18.5k	177.4m	12P	-	158T	A	NO
18	TOR NERINGA	12494t	75	19.0k	168.0m	12P	-	122T	A	LT
19	TOR PETUNIA	32289t	04	22.5k	199.8m	12P	-	280T	AS	DK
20	TOR PRIMULA	32289t	04	22.5k	199.8m	12P	-	280T	AS	DK
21	TOR SCANDIA	33652t	81	19.5k	193.6m	12P	400C	206T	A	SW
22	TOR SELANDIA	24196t	98	21.1k	197.5m	12P	-	206T	A	SW
23	TOR SUECIA	24196t	99	21.1k	197.5m	12P	-	206T	A	DK

AURORA Built at Rauma, Finland as the ARCTURUS for *EFFOA* of Finland and chartered to *Finncarriers*. In 1991 renamed the AURORA. In 2003 chartered to *Scandlines Estonia* to operate between Rostock and Helsinki. During 2004 purchased by *Scandlines Estonia*. In 2005 chartered to *DFDS Tor Line* to operate between Immingham and Zeebrugge.

TANGO Built at Wismar, East Germany. Launched as the RITZBERG for *DDR RoRo* of East Germany, but delivered to *NavRom* of Romania and renamed the TUTOVA; she was used for charter to many operators. In 1996 sold to *Octogon Shipping* of Romania and renamed the OCTOGON 3. In November she was chartered to *East Coast Ferries* and placed on their Hull – Dunkerque service. This service ceased in January 1999. Later in 1999 she was renamed the TANGO; she was engaged in a number of short term charters. In 2002 she entered service with *Ferryways* on their Zeebrugge – Killingholme service. In 2004 chartered to *Channel Freight Ferries* and renamed the CFF SOLENT. In late 2004, chartered to *DFDS Tor Line* to operate on the Gothenburg – Copenhagen – Gdansk. This route ended in February 2005. In March she moved to the Immingham – Zeebrugge service. In April replaced by the TOR CIMBRIA. Charter ended. Renamed the FAST INDEPENDENCE

TOR ANGLIA Built by Shiffswerft u Maschinenfabrik Paul Lindenau, Kiel, Germany as the MERZARIO GALLIA and chartered to *Merzario Line* of Italy for services between Italy and Saudi Arabia. In 1981 she was chartered to *Wilhelmsen*, renamed the TANA and used between USA and West Africa. In 1983 she was chartered to *Salenia AB* of Sweden and renamed the NORDIC WASA. In 1987 she had a brief period on charter to *Atlantic Marine* as the AFRICAN GATEWAY and in 1988 she was sold to *Tor Line* and renamed the TOR ANGLIA. In 1989 an additional deck was added. In recent years she operated on the Gothenburg – Ghent service but in late 1998 she was switched to the Immingham – Rotterdam service. In 2001 transferred back to the Gothenburg – Ghent service. In January 2003 chartered to the *British MoD* for three months. Later chartered to the *Danish MoD* for 36 months for NATO service but can be used on other short-term charters if not required. During 2004 she was chartered to *Van-Uden Ro-Ro* for their service from Northern Europe to the Western Mediterranean. In February 2005 briefly chartered to *Cobelfret Ferries*.

TOR BALTICA Built by Hyundai Heavy Industries, Ulsan, South Korea as the ELK for *Stena Rederi* of Sweden and chartered to *P&O Ferrymasters* for use on services from Middlesbrough to Gothenburg and Helsingborg. Purchased by *P&O* in 1981 and lengthened in 1986; she was managed by *P&O North Sea Ferries Ltd*. In 2001 she was sold to *DFDS Tor Line*, who took over management of the vessel. *P&O Ferrymasters'* services ceased in May 2001. She was renamed the TOR BALTICA and transferred to the *DFDS Tor Line* Gothenburg – Harwich route. In 2005 briefly transferred to the new Immingham – Zeebrugge route before moving to the Fredericia – Copenhagen – Klaìpeda service.

TOR BEGONIA Built by Flensburger Schiffbau-Gesellschaft, Flensburg, Germany for *DFDS Tor Line*. Operates on the Gothenburg – Immingham route.

TOR BELGIA Built by Société Metallurgique & Navale Dunkerque-Normandie, Dunkerque, France as the VILLE DU HAVRE for *Société Française de Transports Maritimes* of France. Between 1979 and 1981 she was chartered to *Foss Line*, renamed the FOSS HAVRE and operated between Europe and the Middle East. In 1987 she was renamed the KAMINA. In 1990 she was chartered to *Maersk Line* of Denmark, renamed the MAERSK KENT and used on *Kent Line* services between Dartford and Zeebrugge. In 1992 she was chartered to and later purchased by *Tor Line*, placed on the Gothenburg

– Immingham route and renamed the TOR BRITANNIA. In 1994 she was lengthened by 23.7m. In 1999 she was renamed the TOR BELGIA and was later transferred to the Gothenburg – Brevik – Ghent route. In 2003 she was sold to Norwegian interests and chartered back for four years. In 2004 moved to the new Gothenburg – Copenhagen – Gdansk route. Later chartered to *Trasmediterranea* to operate between Spain and the Canary Islands.

TOR BRITANNIA Built by Fincantieri Cantieri Navali Italiani, Ancona, Italy for *DFDS Tor Line*. Operated on the Gothenburg – Immingham route until 2004 when she was transferred to the Esbjerg – Immingham route.

TOR CIMBRIA Built by Frederikshavn Værft A/S, Frederikshavn, Denmark. Launched as the MERCANDIAN EXPRESS II and immediately bare-boat chartered to *DFDS* for their North Sea freight services, being renamed the DANA CIMBRIA. Purchased by *DFDS* in 1989. Until 1996, generally used on Immingham and North Shields – Esbjerg services; between 1996 and 1998 she operated between Immingham and Esbjerg. In 1998 she was transferred to the Immingham – Cuxhaven service. In 2001 renamed the TOR CIMBRIA. In 2002 sold to Norwegian interests and chartered back to *DFDS Tor Line*. In 2004 sub-chartered to *HUAL* and used in a trade car carrying role between Brazil and South Africa. In April 2005 placed on the Zeebrugge – Immingham route.

TOR DANIA Built by Société Metallurgique & Navale Dunkerque-Normandie, Dunkerque, France as the VILLE DE DUNKERQUE for *Société Française de Transports Maritimes* of France. Between 1979 and 1981 she was chartered to *Foss Line*, renamed the FOSS DUNKERQUE and operated between Europe and the Middle East. In 1986 she was chartered to *Grimaldi* of Italy and renamed the G AND C EXPRESS. In 1988 she was briefly chartered to *Elbe-Humber RoLine* and renamed the RAILRO. She was then chartered to *DFDS* where she was renamed the DANIA HAFNIA. The following year she was chartered to *Maersk Line* of Denmark, renamed the MAERSK ESSEX and used on *Kent Line* services between Dartford and Zeebrugge. In 1992 she was chartered to and later purchased by *DFDS* and renamed the TOR DANIA. In 1993 she was renamed the BRIT DANIA but later in the year reverted to her original name. She was generally used on the Harwich – Esbjerg service, working in consort with the passenger ferry DANIA ANGLIA (see *DFDS Seaways*). In 1994 she was lengthened by 23.7m. and chartered to *Tor Line* and placed on the Gothenburg – Immingham route. She now operates on the Gothenburg – Brevik – Ghent route. In 2003 she was sold to Norwegian interests and chartered back for four years. In 2004 moved to the Immingham – Cuxhaven route.

TOR FLANDRIA Built by Kockums Varvet, Malmö, Sweden as the FINNCLIPPER for the *Johansson Group* of Sweden and chartered out. In 1983 she was sold to *Zenit Shipping* and renamed the ZENIT CLIPPER. She was chartered to *Foss Line* and used on services between Northern Europe and the Middle East. In 1986 she was sold to *Crowley American Transport* of the USA and chartered to the US Military. She was renamed the AMERICAN FALCON and used for military transport purposes across the world. In 1998 sold to *Stena Rederi* and was renamed the STENA PARTNER. She was then chartered to *Tor Line* and renamed the TOR FLANDRIA; part of her charter conditions are that she be purchased at the end of the five year charter period; however, she was purchased in 2001. She is normally used on the Gothenburg – Brevik – Ghent route. In 2002 she was sold to *Norwegian Scandinavian Ro/Ro KS* and chartered back for 5.5 years.

TOR FREESIA Built by Flensburger Schiffbau-Gesellschaft, Flensburg, Germany for *DFDS Tor Line*. Operates on the Gothenburg – Immingham route.

TOR FUTURA Built by C N Visentini di Visentini Francesco & C, Donada, Italy as the DANA FUTURA for *DFDS*. In 2001 she was renamed the TOR FUTURA. Initially operated mainly between Esbjerg and Harwich, but latterly operated mainly between Esbjerg and Immingham. In 2004 chartered to *Toll Shipping* of Australia. Later chartered to the *Danish MoD* for 5.5 years. However, when not required for military service she has been chartered to other operators such as *P&O Ferries* and *Cobelfret Ferries*.

TOR HOLLANDIA Built by A/S Framnæs Mekaniske Verksted, Sandefjord, Norway as the TOR DANIA for charter to *Tor Line*. In 1975 she was chartered to *Salenrederierna* for service in the Middle East and renamed the BANDAR ABBAS EXPRESS. In 1977 she was lengthened and, in 1978, returned to *Tor Line* and resumed the name TOR DANIA. Purchased by *Tor Line* in 1986. In 1992 she was renamed the TOR DAN and in 1993 the TOR HOLLANDIA. She was usually used on the Immingham

– Rotterdam service. In 1999 transferred to the Norway – UK/Netherlands services (Netherlands services ceased in 2001). In 2002 moved to the Norway – Immingham route. After a period operating between Cuxhaven and Immingham, she now operates on the Norway route again.

TOR HUMBRIA Built by Oskarshamns Varvs Ab, Oskarshamn, Sweden as the EMIRATES EXPRESS for *A/S Skarhamns Oljetransport* of Norway and chartered to *Mideastcargo* for services between Europe and the Middle East. In 1981 chartered to *OT West Africa Line* for services between Europe and West Africa and renamed the ABUJA EXPRESS. In 1983 chartered to *Foss Line*, renamed the FOSSEAGLE and returned to Middle East service. In 1985 she was renamed the FINNEAGLE, chartered briefly to *Finncarriers* and then to *Fred. Olsen Lines*. In 1987 they purchased her and renamed her the BORAC. In 1999 purchased by *DFDS Tor Line* and renamed the TOR HUMBRIA. In 2000 she was chartered to *Costa Container Lines spa* of Italy, operating between Savano and Catania. This service ended in early 2001 and she was then chartered to *CoTuNav* of Tunisia. Returned in April 2001. In 2003 sold to Norwegian interests and chartered back. She currently operates on the Rotterdam – Immingham service.

TOR MAGNOLIA Built by Flensburger Schiffbau-Gesellschaft, Flensburg, Germany. Initially operated on the Gothenburg – Immingham route. In 2005 moved to the Gothenburg – Ghent route.

TOR MAXIMA Built by Hitachi Zosen, Osaka Works, Sakai, Japan as the DANA MAXIMA for *DFDS* for their North Sea services. Until 1996, generally used on the Esbjerg – Grimsby and North Shields services. In summer 1995 she was lengthened to increase trailer capacity. In December 2000 she was renamed the TOR MAXIMA. In December 2001 sold to *Per Sand* of Norway and chartered back for 3 years. She currently operates between Cuxhaven and Immingham.

TOR MINERVA Built by Oskarshamns Varvs Ab, Oskarshamn, Sweden as the BANDAR ABBAS EXPRESS for *A/S Skarhamns Oljetransport* of Norway and chartered out. In 1980 renamed the SAUDI EXPRESS. During the early eighties she undertook a number of charters including *Mideastcargo* for services between Europe and the Middle East, *Atlanticargo* for services from Europe to USA and Mexico and *OT West Africa Line* from Europe to West Africa. In 1983 she was chartered to *Ignazio Messina* of Italy, renamed the JOLLY AVORIO and used on services from Italy to the Middle East. In 1986 this charter ended and she briefly reverted to the name the SAUDI EXPRESS before being chartered again to *OT West Africa Line* and renamed the KARAWA. In 1987 she was sold to *Fred. Olsen Lines* who renamed her the BORACAY; she operated between Norway and Northern Europe. In 1998 she was sold to *DFDS*, renamed the DANA MINERVA and placed on the Esbjerg – Immingham route. In 2001 she was renamed the TOR MINERVA. She now operates between Rotterdam and Immingham.

TOR NERINGA Built by Ankerløkken Verft, Florø, Norway as the BALDUIN for *Fred. Olsen Lines*. In 1999 purchased by *DFDS Tor Line* and renamed the TOR NORVEGIA. Initially used on Norway – UK/Netherlands services; in 2001 moved to the Fredericia – Copenhagen – Klaìpeda (Lithuania) service. In December 2001 sold to *Lisco Baltic Service* of Lithuania, renamed the TOR NERINGA and chartered back to *DFDS Tor Line*. In 2004 moved to the Gothenburg – Rotterdam/Harwich route. In 2005 moved to the new Immingham – Zeebrugge route. In March 2005 chartered to *Cobelfret Ferries* to operate between Rotterdam and Purfleet; charter expected to end June 2005 and she will move to the new Gothenburg - Tilbury service.

TOR PETUNIA Built by Flensburger Schiffbau-Gesellschaft, Flensburg, Germany for *DFDS Tor Line*. Operates on the Gothenburg – Immingham route.

TOR PRIMULA Built by Flensburger Schiffbau-Gesellschaft, Flensburg, Germany. Initially operated on the Gothenburg – Immingham route. Later in 2004 moved to the Gothenburg – Ghent route.

TOR SCANDIA Built by Kockums Varvet, Malmö, Sweden as the KUWAIT EXPRESS for the *Johansson Group* of Sweden and chartered to *NYK Line* of Japan for services between Japan and the Arabian Gulf. In 1983 she was sold to *Zenit Shipping* and renamed the ZENIT EXPRESS. She was chartered to *Foss Line* and used on services between Northern Europe and the Middle East. In 1984 she was sold to *Crowley American Transport* of the USA and chartered to the US Military. She was reamed the AMERICAN CONDOR and used for military transport purposes across the world. In 1998 she was sold to *Stena Rederi* and renamed the STENA PORTER. Later she was sold to *Tor Line* and renamed the TOR SCANDIA. She is used on the Gothenburg – Brevik – Ghent route. In 2002 she was sold to

Norwegian Scandinavian Ro/Ro KS and chartered back for 5.5 years.

TOR SELANDIA Built by Fincantieri Cantieri Navali Italiani, Ancona, Italy for *DFDS Tor Line*. Operated on the Gothenburg – Immingham route until 2004 when she was moved to the Gothenburg – Ghent route. In 2005 she moved to the Gothenburg – Harwich (Tilbury as from July 2005) route.

TOR SUECIA Built by Fincantieri Cantieri Navali Italiani, Ancona, Italy for *DFDS Tor Line*. Operated on the Gothenburg – Immingham route until 2004 when she was transferred to the Esbjerg – Immingham route. Later transferred to the Danish flag.

Planned Charter

24	TOR BELLONA	22748t	80	18.0k	178.5m	12P	-	155T	A	UK

TOR BELLONA Built by Kawasaki Heavy Industries, Sakaide, Japan as the ZHANG JIA KOU, a deep sea ro-ro/container ship for *China Ocean Shipping Company* of the People's Republic of China for service between the USA, Australia and New Zealand. In 1999, purchased by *Jacobs Holdings*. After delivery, she was converted in Nantong, China to short sea ro-ro specification, including the fitting of a stern ramp (replacing the quarter ramp) and luxury accommodation for 12 drivers. Rebuilt as the DART 10. On completion of rebuilding, chartered to *Sudcargos* of France and renamed the MONT VENTOUX and operated between France and North Africa. In December 2000 the charter ended. In January 2001 she briefly ran on the Dartford – Zeebrugge route in place of the DART 8. After two brief charters to the *British MoD* she was refitted and renamed the DART 10. She then entered long term charter with the *British MoD*. Returned to *Dart Line* in August 2003 and operated between Dartford and Zeebrugge for about three weeks before going on another *MoD* charter to Iraq. She was then chartered to *CETAM* of France and renamed the MASSILIA. In early 2005 sold to a limited partnership under control of the *Eidsiva Rederi* of Norway; continued to be chartered to *CETAM*. In autumn 2005 to be chartered to *DFDS Tor Line* and renamed the TOR BELLONA. Deployment not yet decided.

Under Construction

25	TOR FICARIA	32400t	06	22.5k	199.8m	12P	-	280T	AS	DK

TOR FICARIA Under construction at Flensburger Schiffbau-Gesellschaft, Flensburg, Germany for *DFDS Tor Line*. Another 'flower class' vessel.

FERRYWAYS

THE COMPANY *Ferryways nv* is a Belgian company.

MANAGEMENT Managing Director J Dewilde, **Marketing Manager** Filip Olde Bijvank, **Commercial Manager** Leon van Eecke.

ADDRESS ***Ostend*** Esplanadestraat 10, 8400 Ostend, Belgium, ***Ipswich*** West Bank Terminal, Wherstead Road, Ipswich IP2 8NB.

TELEPHONE Administration & Reservations ***Ostend*** +32 (0)59 34 22 26, ***Ipswich*** +44 (0)1473 696200.

FAX Administration & Reservations ***Ostend*** +32 (0)59 34 22 29, ***Ipswich*** +44 (0)1473 696201.

INTERNET Email ***Ostend*** info@ferryways.com ***Ipswich*** **Website** www.ferryways.be *(English)*

ROUTE OPERATED Ostend *(11.00 Tue -Fri, 15.00 Sat, 17.00 Mon-Fri, 22.00 Sun-Fri,)* – Ipswich *(dep: 03.00 Tue-Sat, 09.30 Tue-Sat, 21.00 Sun, 22.00 Mon-Fri)* (7 hrs; *(1,3,6)*; 3 per day), Ostend *(dep: 02.00 Tue-Sat, 22.00 Sat)* – Immingham (Killingholme) *(dep: 23.59 Mon-Sat)* (14/19 hrs; *(2,4,5)*; 1 per day, Ostend *(dep: 18.00 Mon, Wed, Fri, 23.59 Mon-Sat (via Killingholme))* – Immingham (Docks) *(dep: 17.00 Tue, 17.30 Thu, 19.30 Mon-Sat (via Killingholme))* (14/19 hrs; *(2,4,5)*; 1 per day.

1	ANGLIAN WAY	7635t	78	15.0k	141.3m	12P	55C	84T	A	PA
2	CALIBUR	9963t	76	16.0k	126.4m	12P	-	95T	A	MA
3	FLANDERS WAY	7628t	77	16.0k	141.3m	12P	55C	84T	A	PA

4	HUMBER WAY	9963t	76	16.0k	132.5m	12P	-	96T	A	MA
5	IPSWICH WAY	6568t	80	15.0k	136.0m	12P	-	84T	A	PA
6	OSTEND WAY	6568t	80	15.0k	136.1m	12P	-	84T	A	PA

ANGLIAN WAY Built by Rickmers Werft GmbH, Bremerhaven, Germany as the THOMAS WEHR for *Wehr Transport* of Germany as THOMAS WEHR but on delivery chartered to *Wacro Line* and renamed the WACRO EXPRESS. In 1978 charter ended and she was renamed the THOMAS WEHR. Over the next few years she was chartered to several operators. In 1982 she was chartered to *Tor Lloyd* (later *Tor Line*) for North Sea service and renamed the TOR NEERLANDIA. In 1985 the charter was transferred to *DFDS* and she was renamed the DANA GERMANIA. This charter terminated in 1985 and she resumed her original name. In early 1986 she was chartered to *North Sea Ferries* for their Hull – Zeebrugge service. This charter ended in summer 1987. Subsequent charters included *Cobelfret* and *Elbe-Humber RoLine* and a twelve month period with *North Sea Ferries* again – this time on the Hull – Rotterdam and Teesport – Zeebrugge routes. In 1993 she was renamed the MANA, then the SANTA MARIA and finally chartered to *TT-Line* and renamed the FULDATAL. 1994 she was chartered to *Horn Line* for service between Europe and the Caribbean and renamed the HORNLINK. Later that year she was chartered to *P&O European Ferries* for the Portsmouth – Le Havre freight service and resumed the name THOMAS WEHR. In late 1995 transferred to the Felixstowe – Zeebrugge freight service. In autumn 1999 the charter was ended. In 2000 she was chartered to *Ferryways*. In 2001 she was purchased by *Ferryways* and renamed the ANGLIAN WAY. Generally operates between Ostend and Ipswich.

CALIBUR Built by Trosvik Verksted A/S, Brevik, Norway as the SEASPEED DANA for *Seaspeed Ferries* of Greece and chartered to *Fred. Olsen Seaspeed Ferries* for service in the Middle East. In 1979 chartered to *Roto Line* and used on services between Sweden and the UK. , renamed the DANA In 1983 she was sold to *Stena Line* and embarked on a number of charters, mainly in the Caribbean. In 1990 she was chartered to *CoTuNav* of Tunisia, renamed the SALAH L and used on services from Tunisia to Southern Europe. In 1993 she was chartered to *Olympic Ferries* of Greece, renamed the SENATOR, operating between Greece and Italy. In 1995 she was sold to *Exxtor Ferries* and renamed the EXCALIBUR. She operated between Immingham and Rotterdam. When the service was taken over by *Cobelfret Ferries* in 1997, the charter was ended. She was sold to *Kaliningrad Key Kine* of Malta, renamed the CALIBUR and placed on their Køge (Denmark) – Aabenraa (Denmark) – Kaliningrad service. In 1998 she was sold to *V Ships* of Monaco and placed on the charter market, undertaking a number of short charters. In 2003 she was chartered to *Ferryways* and placed on their Ostend – Killingholme service.

FLANDERS WAY Built by Rickmers Werft GmbH, Bremerhaven, Germany as the GABRIELE WEHR for *Wehr Transport* of Germany and chartered to several operators. In 1982, chartered to *Tor Lloyd* (later *Tor Line*) for North Sea service and renamed the TOR ANGLIA. This charter terminated in 1985 when she resumed her original name and, in early 1986, she was chartered to *North Sea Ferries* for their Hull – Zeebrugge service. This charter ended in summer 1987 when the lengthened NORLAND and NORSTAR entered service. Subsequent charters included *Kent Line* and *Brittany Ferries*. In 1989 she was chartered to *P&O European Ferries* for the Portsmouth – Le Havre freight service. Her charter was terminated following the transfer of the EUROPEAN TRADER to the route in late 1992 but in 1993 it was renewed, following the transfer of the EUROPEAN CLEARWAY (now the EUROPEAN PATHFINDER) to *Pandoro*. In 1996, she was transferred to the Felixstowe – Zeebrugge service. In autumn 1999 the charter was ended. In 2000 she was chartered to *Ferryways*. In 2001 she was purchased by *Ferryways* and renamed the FLANDERS WAY. Generally operates between Ostend and Ipswich.

HUMBER WAY Built by Porsgrund Mekaniske Verksted, Porsgrunn, Norway, as the SEASPEED DORA for *Seaspeed Ferries* of Greece and chartered to *Fred. Olsen Seaspeed Ferries* for service in the Middle East. In renamed the INGER EXPRESS and in 1979 chartered to *Roto Line*, used on services between Sweden and the UK. In 1981 she was sold to *Cobelfret*, renamed the MARCEL C and used on their services between the UK and Belgium. In 1989 she was sold to *Bassro Star A/S* of the Bahamas and renamed the BASSRO STAR. She embarked on a series of charters including *Belfast Freight Ferries*, *North Sea Ferries* and a number of other operators. In 1995 she was sold to *Exxtor Ferries* and renamed the ENDEAVOUR. In 1997 she was sold to *Cobelfret*, when they took over the *Exxtor Ferries'* service, and renamed the MARCELINE. She continued to operate between

Immingham and Rotterdam. In 1998 she was sold to *East West Mediterranean Ltd* of Malta (trading as *Van Uden RoRo*) and renamed the BEATRIXHAVEN. She operated between Europe and the Middle East plus undertaking a number of short term charters. In 2003 she was chartered to *Ferryways* and later renamed the HUMBER WAY. Generally operates between Ipswich and Killingholme.

IPSWICH WAY Built by Karlskronavarvet AB, Karslkrona, Sweden as the BALDER DONA for *Dag Engström Rederi* of Sweden and undertook a number of charters in the Caribbean and Mediterranean. In 1984 she was renamed the RODONA and chartered to *Seaboard Shipping* of the USA and used on Caribbean services. In 1987 she was chartered to the *Ford Motor Company* for conveyance of privately owned trailers between Dagenham and Zeebrugge. In 1995 *Cobelfret Ferries* took over the operation of this service and she was used on both the Purfleet – Zeebrugge and Dagenham – Zeebrugge services. In 1999 she was chartered to *P&O North Sea Ferries* to operate between Felixstowe and Zeebrugge. In 2002 this service ceased and she was chartered to *Ferryways*. In 2003 purchased by *Ferryways* and renamed the IPSWICH WAY. Generally operates between Ostend and Ipswich.

OSTEND WAY Built by Karlskronavarvet AB, Karlskrona, Sweden as the BALDER VINGA for *Dag Engström Rederi* of Sweden and undertook a number of charters in the Caribbean and Mediterranean. In 1984 she was renamed the ROVINGA and chartered to *Seaboard Shipping* of the USA and used on Caribbean services. In 1985 she was renamed the AZUA. In 1987 she briefly reverted to the name ROVINGA before being renamed the SAPPHIRE and chartered to the *Ford Motor Company* for conveyance of privately owned trailers between Dagenham and Zeebrugge. Since 1995, as the IPSWICH WAY. Normally operates on the Ipswich service. In 2003 purchased by *Ferryways* and renamed the OSTEND WAY.

FINNLINES

THE COMPANY *Finnlines PLC* is a Finnish private sector company. Services to the UK are marketed by *Finnlines UK Ltd*, a British private sector company. From 1st January 2001, *Finncarriers* was merged into the parent company, trading as *Finnlines Cargo Service*.

MANAGEMENT *Finnlines* President Antti Lageroos, **Vice-President** Simo Airas.

ADDRESS *Finnlines* PO Box 197, Salmisaarenkatu 1, 00181 Helsinki, Finland, ***Finnlines UK Ltd*** 8 Heron Quay, London E14 4JB.

TELEPHONE Administration & Reservations *Finnlines* +358 (0)10 34350, ***Finnlines UK Ltd*** +44 (0)20 7519 7300.

FAX Administration *Finnlines* +358 (0)10 3435200, ***Finnlines UK Ltd*** +44 (0)20 7536 0255.

INTERNET Email *Finnlines* info@finnlines.fi ***Finnlines UK Ltd*** london@finnlines.co.uk

Websites www.finnlines.fi *(English, Finnish)*

ROUTES OPERATED *Circuit 1* Helsinki *(dep: Tue)* – St Petersburg *(dep: Wed)* – Hamina *(dep: Thu)* – Helsinki *(dep: Fri)* – Zeebrugge *(dep: Mon)* – Felixstowe *(dep: Tue)* – Amsterdam *(dep: Wed)* – Antwerp *(dep: Thu)* – Baltijsk (Kaliningrad) *(dep: Sun)* – Helsinki *(arr: Mon) (19)*, ***Circuit 2*** Helsinki *(dep: Fri)* – Zeebrugge *(dep: Mon)* – Felixstowe *(dep: Tue)* – Amsterdam *(dep: Wed)* – Antwerp *(dep Thu)* – Helsinki *(arr: Mon)* – Hamina *(arr: Tue, dep: Thu)* – Helsinki *(arr: Fri) (3)* (Note: the two fortnightly services give a weekly departure from Helsinki and Hamina to Zeebrugge, Felixstowe, Amsterdam and Antwerp and fortnightly departures from St Petersburg and Baltijsk), ***Circuit 3*** Hamina *(dep: Week 1 – Wed, Week 2 -Fri, Week 3 -No sailing)* – Felixstowe *(dep: Week 1 – Sun, Week 2 – No sailing, Week 3 – Thu)* (three week rolling schedule) (*17*), ***Circuit 4*** Kemi *(dep: 16.00 Mon)* – Oulu *(arr: 06.00 Tue, dep: 22.00 Tue)* – Felixstowe *(arr: 08.00 Sat, dep: 16.00 Sat)* – Antwerp *(arr: 08.00 Sun, dep: 16.00 Sun)* – Helsinki *(arr: 07.00 Wed, dep 11.00 Wed)* – Kemi *(arr: 16.00 Thu, dep: 23.00 Thu)* – Oulu *(arr: 06.00 Fri, dep: 22.00 Fri)* – Antwerp *(arr: 14.00 Tue, dep: 22.00 Tue)* – Felixstowe *(arr: 08.00 Wed, dep: 14.00 Wed)* – Helsinki *(arr: 07.00 Sat, dep: 12.00 Sat)* – Turku *(arr: 08.00 Sun, 11.00 Sun)* – Kemi *(arr: 08.00 Mon)(4,5,6)*; three ships operate on a three week cycle which gives two sailings per week between Finland and UK/Benelux), ***Circuit 5*** Hamina *(dep: 23.00 Thu)* – Helsinki *(arr: 07.00 Fri, dep: 15.00 Fri)* – Hull *(arr: 07.00 Mon, dep: 18.00 Mon)* – Hamina *(arr: 12.00 Thu)* (3 days; *(11)*; 1 per week), ***Circuit 6*** Rauma *(dep: 20.00 Wed)* – Hull *(arr:*

07.00 Sat, dep: 20.00 Sat) – Helsinki *(arr: 08.00 Tue, dep: 15.00 Tue)* – Rauma *(arr: 07.00 Wed)*, (3 days; *(15)*; 1 per week).

In view of the fact that ships are liable to be transferred between routes, the following is a list of all *Finnlines Cargo* ro-ro vessels, including those which currently do not serve the UK. Ro-pax vessels (none of which normally serve the UK) are listed in Section 6.

1	ANTARES	5989t	88	20.3k	157.6m	18P	-	154T	A	NO
2	ASTREA	7380t	91	15.0k	129.1m	0P	-	60T	A	FI
3	BALTICA	21224t	90	19.0k	157.7m	0P	-	163T	A	CY
4	BIRKA CARRIER	12251t	98	20.0k	155.5m	12P	-	124T	A2	FI
5	BIRKA EXPRESS	12251t	97	20.0k	154.5m	12P	-	124T	A2	FI
6	BIRKA TRADER	12251t	98	20.0k	154.5m	12P	-	124T	A2	FI
7	FINNBIRCH	15396t	78	17.0k	155.9m	0P	-	155T	A	SW
8	FINNFOREST	15525t	78	17.0k	155.9m	0P	-	155T	A	SW
9	FINNHAWK	11530t	01	20.0k	162.2m	12P	-	140T	A	NO
10	FINNKRAFT	11530t	00	20.0k	162.2m	12P	-	140T	A	UK
11	FINNMASTER	11530t	00	20.0k	162.2m	12P	-	140T	A	UK
12	FINNMILL	11400t	02	20.0k	184.8m	12P	-	230T	A	UK
13	FINNOAK	7953t	91	16.5k	139.5m	0P	-	94T	A	FI
14	FINNPULP	11400t	02	20.0k	184.8m	12P	-	230T	A	UK
15	FINNREEL	11530t	00	20.0k	162.2m	12P	-	140T	A	UK
16	FORTE	3998t	89	15.2k	90/8m	0P	-	20T	AS	NL
17	KADUNA	3166t	72	16.5k	113.5m	12P	-	70T	A	PA
18	LARGO	3998	90	15,0k	90.5m	0P		20T	AS	NL
19	MERCHANT	21195t	82	17.0k	154.9m	12P	-	160T	A	FI
20	MIRANDA	10471t	98	22.0k	153.5m	12P	-	112T	A	FI
21	MISTRAL	10471t	98	22.0k	153.5m	12P	-	112T	A	FI
22	NORCLIFF	8407t	94	14.5k	125.2m	0P	-	78T	A	NO
23	POLARIS	7944t	88	14.7k	122.0m	0P	-	38T	A	GY
24	RIDER	20077t	84	18.9k	186.5m	12P	-	140T	A	FI
25	RUNNER	20729t	90	18.9k	189.7m	12P	-	144t	A	FI
26	VASALAND	20203t	84	14.0k	155.0m	0P	-	160T	A	SW

ANTARES Built by Gdansk Shipyard, Gdansk, Poland as the FINNFORREST for *Neste* of Finland and chartered to *Finncarriers*. In 1988 renamed the ANTARES. In 2002 chartered to *Stena Line* to operate between Harwich and Rotterdam. In 2003 chartered to the *British MoD* for service to the Gulf. Currently operates on the Oulu – Kemi – Turku – Lübeck route.

ASTREA Built by Tangen Verft Kragerø A/S, Tomrefjord, Norway for *Finncarriers*. Operates between Finland and Spain – Portugal via Antwerp.

BALTICA Built by Hyundai Heavy Industries, Ulsan, South Korea as the AHLERS BALTIC the for *Ahlers Line* and chartered to *Finncarriers*. In 1995 acquired by *Poseidon Schiffahrt AG* of Germany and renamed the TRANSBALTICA. She continued to be chartered to *Finncarriers* and was acquired by them when they purchased *Poseidon Schiffahrt AG* (now *Finnlines Deutschland AG*) in 1997. In 2003 sold to Norwegian interests and chartered back. She was renamed the BALTICA. Currently operating on the Helsinki – Hamina – Felixstowe – Zeebrugge –Antwerp – Amsterdam service.

BIRKA CARRIER, BIRKA EXPRESS, BIRKA TRADER Built by Fosen Mekaniske Verksteder A/S, Rissa, Norway as the UNITED CARRIER, UNITED EXPRESS and UNITED TRADER for *United Shipping* (a subsidiary of *Birka Shipping*) of Finland and chartered to *Transfennica*. During 2000 they were used on their Kemi – Oulu – Antwerp – Felixstowe service. In 2001 the route was transferred to *Finnlines* and vessels used sub-chartered to them. In 2002 *United Shipping* was renamed *Birka Cargo* and the ships were renamed the BIRKA CARRIER, BIRKA EXPRESS and BIRKA TRADER.

FINNBIRCH Laid down by Hyundai Heavy Industries, Ulsan, South Korea as the STENA PROSPER and completed as the ATLANTIC PROSPER for *Stena Rederi* and chartered to *ACL* of Great Britain for service between Britain and Canada. In 1981 chartered to *Merzario Line* of Italy for services between Italy and Saudi Arabia and renamed, initially, the STENA IONIA and then the MERZARIO IONIA. In 1982 she reverted to the name STENA IONIA and was chartered to *OT West Africa Line* for services between Europe and Nigeria. In 1985 she was renamed the STENA GOTHICA and used on *Stena Portlink* services. In 1988 she was chartered to *Bore Line* of Finland and renamed the BORE GOTHICA. In 1992 chartered to *Finncarriers*. In 1996 renamed the FINNBIRCH. In 1997 she began operating a service between Hull and Zeebrugge on charter to *P&O North Sea Ferries* in the course of her normal two week circuit from Finland. This ceased in 1999. In 2000 transferred to the Helsinki – Århus service.

FINNFOREST Laid down by Hyundai Heavy Industries, Ulsan, South Korea as the STENA PROJECT and completed as ATLANTIC PROJECT for *Stena Rederi* and chartered to *ACL* (see above). In 1981 chartered to *Merzario Line* of Italy for services between Italy and Saudi Arabia and renamed the MERZARIO HISPANIA. In 1983 returned to *Stena Line* and renamed the STENA HISPANIA. In 1984 chartered to *Kotka Line* of Finland, renamed the KOTKA VIOLET and used on their services between Finland, UK and West Africa. This charter ended in 1985 and she was again named the STENA HISPANIA. In 1986 she was renamed the STENA BRITANNICA and used on *Stena Portlink* (later *Stena Tor Line*) service between Sweden and Britain. In 1988 she was chartered to *Bore Line* of Finland, renamed the BORE BRITANNICA and used on services between Finland and Britain. In 1992 chartered to *Finncarriers*. In 1997 renamed the FINNFOREST. In 1997 she began operating a service between Hull and Zeebrugge on charter to *P&O North Sea Ferries* in the course of her normal two week circuit from Finland. This ceased in 1999. In 2000 transferred to the Helsinki – Århus service.

FINNHAWK Built by Jinling Shipyard, Nanjing, China for *Finnlines*. Currently alternates week about on the Rauma – Lübeck – Helsinki – Travemünde and Helsinki – Copenhagen – Oslo – Helsingborg routes.

FINNKRAFT Built by Jinling Shipyard, Nanjing, China for *Forest Terminals* and chartered to *Finncarriers*. Currently alternates week about on the Rauma – Lübeck – Helsinki – Travemünde and Helsinki – Copenhagen – Oslo – Helsingborg routes.

FINNMASTER Built by Jinling Shipyard, Nanjing, China for *Forest Terminals* and chartered to *Finncarriers*. Operates on Helsinki – Hamina – Hull route.

FINNMILL Built by Jinling Shipyard, Nanjing, China for *Forest Terminals* and chartered to *Finnlines*. Currently used on the Helsinki – Kotka – Rauma – Lübeck route.

FINNOAK Built by Fosen Mekaniske Verksteder A/S, Rissa, Norway as the AHTELA for *Holming Shipping* of Finland and chartered to *Transfennica*. In 1997 renamed the FINNOAK and chartered to *Finncarriers*. Used on the *Polfin Line* service between Helsinki – Kotka and Gdynia – Szczecin (joint with *Euroafrica Shipping* of Poland).

FINNPULP Built by Jinling Shipyard, Nanjing, China for *Forest Terminals* and chartered to *Finnlines*. Currently used on the Helsinki – Kotka – Rauma – Lübeck service.

FINNREEL Launched as the FINNMAID but renamed before delivery. Built by Jinling Shipyard, Nanjing, China for *Forest Terminals* and chartered to *Finncarriers*. Currently operating on the Hamina – Rauma – Hull service.

FORTE Built by Bodewes Shipyards, Hoogezand, Netherlands for Dutch interests and chartered to *Bror Husell* of Finland. Sub-chartered to *Transfennica*, operating between Finland and the UK. In 1994 sold to *B&N Nordsjöfrakt* and chartered to *Finnlines*. Operates between Kemi and Bilbao.

KADUNA Built as the BORE IX by Rauma-Reopla Oy, Rauma, Finland for *Bore Line* of Finland. In 1977 sold to *VVB Schiffbau* of East Germany and renamed the ASCHBERG. In 1990 transferred to DSR and in 1994 to *Euroseabridge*. In 1996 sold to *E H Harms* of Germany and renamed the FEEDERCADET. Used for car transport. In 1994 sold to *Barbura Line* of Panama. In 2004 renamed the KADUNA and later chartered to *Finnlines* to operate between Helsinki, Hamina and Felixstowe undertaking two round trips every three weeks.

LARGO Built by Bodewes Shipyards, Hoogezand, Netherlands for Dutch interests and chartered to *Bror Husell* of Finland. Sub-chartered to *Transfennica*, operating between Finland and the UK. In

1994 sold to *B&N Nordsjöfrakt* and chartered to *Finnlines*. Operates between Kemi and Bilbao.

MERCHANT Built by Rauma Repola Oy, Rauma, Finland as the FINNMERCHANT for *Finnlines*. In 2003 sold to subsidiary *Nordö-Link* and chartered back. Later sold to *Finn Ro Ro* of Norway and renamed the MERCHANT; charter continued. Currently operates on the Helsinki – Hamina- St Petersburg – Felixstowe – Amsterdam – Zeebrugge – Baltijsk service.

MIRANDA Built by J J Sietas KG, Hamburg, Germany for *Godby Shipping A/S* of Finland. Initially chartered to *Transfennica*. In 2000 she was chartered to *Finnlines*. Currently used on a Helsinki – Hamina – Zeebrugge service not available for general traffic.

MISTRAL Built by J J Sietas KG, Hamburg, Germany for *Godby Shipping AB* of Finland. Chartered to *Transfennica*. In 2003 chartered to *UPM Kymmene Oy* of Finland and operated between Rauma and Santander. In 2005 chartered to *Finnlines*. Currently used on a Helsinki – Hamina – Zeebrugge service not available for general traffic.

NORCLIFF Built by Brodogradiliste Trogir, Trogir, Croatia for *Sea-Link AB* of Sweden but due to delays order cancelled before completion. On delivery, renamed the BRAVO and chartered to the *Stora Paper Group*. In 1995 chartered to *North Sea Ferries* and renamed the NORCLIFF. She became second vessel on the Middlesbrough – Rotterdam service. The charter ended in 1996 when the service reverted to a single ship operation. She kept her *NSF* name and was chartered to a number of operators including *Finnlines (Finncarriers)*. In 2001 again chartered to *Finnlines* and is currently used on the Finland to Spain – Portugal via Antwerp service.

POLARIS Built by J J Sietas Werft, Hamburg, Germany for *Pohl Shipping* and chartered to Finncarriers. Currently operating between Finland and Spain – Portugal via Antwerp.

RIDER Built by Schichau Seebeckwerft AG, Bremerhaven, Germany as the RAILSHIP II for Railship, later taken over by *Finnlines*. In 2002 renamed the FINNRIDER. In 2003 sold to Norwegian interests and chartered back. In 2004 renamed the RIDER. Operates between Travemünde and Turku.

RUNNER Built by Schichau Seebeckwerft GmbH, Bremerhaven, Germany as the RAILSHIP III for *Railship*, later taken over by Finnlines. In 2002 role changed from rail ferry to truck and trailer ferry; renamed the FINNRUNNER. In 2003 sold to Norwegian interests and chartered back. In 2004 renamed the RUNNER. Operates between Travemünde and Turku.

VASALAND Built by Rauma Repola Oy, Rauma, Finland as the OIHONNA for *Finncarriers*. In 2003 sold to *Stena RoRo*. Later sold to *Imperial RoRo*, chartered back to *Stena RoRo* and then time chartered to *Finnlines*. In 2004 transferred to *TranRussiaExpress*. In 2005 transferred back to *Finnlines*. Operates on the Rauma – Rostock – Lübeck – Travemünde service.

FORELAND SHIPPING

THE COMPANY *Foreland Shipping Limited* (formerly *AWSR Shipping Limited*) is a UK private sector company jointly owned by *Bibby Line, James Fisher, Houlder Hadley* and *Andrew Weir*.

MANAGEMENT Managing Director Max J Gladwyn.

ADDRESS Dexter House, 2 Royal Mint Court, London EC3N 4XX.

TELEPHONE Administration +44 (0)20 7480 4140.

FAX Administration +44 (0)20 7481 9940.

INTERNET Email max.gladwyn@foreland-shipping.co.uk **Website** www.foreland-shipping.co.uk *(English)*

ROUTES OPERATED No routes are operated. The ships are for charter to the UK *Ministry of Defence* for their Strategic Sealift capability. Normally two of the ships are chartered commercially but can be recalled in times of emergency.

1	ANVIL POINT	23235t	03	18.0k	193.0m	12P	-	180T	A	UK
2	BEACHY HEAD	23235t	03	21.0k	193.0m	12P	-	180T	A	UK
3	EDDYSTONE	23235t	02	21.0k	193.0m	12P	-	180T	A	UK

Humber Way *(Miles Cowsill)*

Hascosay *(Miles Cowsill)*

4	HARTLAND POINT	23235t	03	18.0k	193.0m	12P	-	180T	A	UK
5	HURST POINT	23235t	02	18.0k	193.0m	12P	-	180T	A	UK
6	LONGSTONE	23235t	03	21.0k	193.0m	12P	-	180T	A	UK

ANVIL POINT, HARTLAND POINT Built by Harland and Wolff, Belfast, UK for *AWSR Shipping*.

BEACHY HEAD, LONGSTONE Built by Flensburger Schiffbau-Gesellschaft, Flensburg, Germany for *AWSR Shipping*. Currently on charter to *Transfennica* and operating between Hanko (Finland) and Lübeck (Germany).

EDDYSTONE, HURST POINT Built by Flensburger Schiffbau-Gesellschaft, Flensburg, Germany for *AWSR Shipping*.

MANN LINES

THE COMPANY *Mann Lines* is owned by *Mann & Son (London) Ltd* of Great Britain. It replaced in 2001 *ArgoMann Ferry Service*, a joint venture between *Argo Reederei* of Germany and *Mann & Son*.

MANAGEMENT Managing Director Bill Binks.

ADDRESS ***UK*** Mann & Son (London) Ltd, The Naval House, Kings Quay Street, Harwich CO12 3JJ, ***Germany*** Mann Lines GmbH, Birkenstrasse 15, 28195 Bremen.

TELEPHONE Administration & Reservations ***UK*** +44 (0)1255 245200, ***Germany*** +49 (0)421 163850, ***Finland*** +352 (0)22750000, ***Estonia*** +372 (0)679 1450.

FAX Administration & Reservations ***UK*** +44 (0)1255 245219, ***Germany*** +49 (0)421 1638520

INTERNET Email Michael.footitt@mannlines.co.uk **Website** www.mannlines.com *(English, German, Finnish, Estonian)*

ROUTE OPERATED Harwich (Navyard) *(dep: 22.00 Fri)* – Cuxhaven *(arr: 17.00 Sat, dep: 19.00 Sat)* – Paldiski *(arr: 15.00 Mon, dep: 21.00 Mon)* – Turku *(arr: 08.00 Tue, dep: 17.00 Tue)* – Bremerhaven *(arr: 18.00 Thu, dep: 21.00 Thu)* – Harwich *(arr: 16.00 Fri)*; *(1)*; one per week).

1	ESTRADEN	18205t	99	20.0k	162.7m	12P	130C	170T	A	FI

ESTRADEN Built by Aker Finnyards, Rauma, Finland as the ESTRADEN for *Rederi Ab Engship* of Finland and chartered to *ArgoMann*. Later in 1999 renamed the AMAZON. In 2001 charter was taken over by *Mann Lines* and later in the year she resumed the name ESTRADEN.

NORFOLKLINE

THE COMPANY, MANAGEMENT, ADDRESS, TELEPHONE & INTERNET See Section 1.

ROUTES OPERATED Felixstowe *(dep: 00.30 Daily, 06.00 Tue-Fr, 12.00 Daily, 19.00 Mon-Fri)* – Scheveningen *(dep: 07.00 Tue-Fri, 14.30 Daily, 19.30 Mon-Sat, 22.00 Sun, 23.59 Mon-Fri)* (7 hrs; *(1,2,3,4)*; 4 per day) (Note Service likely to move from Scheveningen to Vlaardingen (near Rotterdam) in early 2006), Immingham – Esbjerg (6 per week), Harwich – Esbjerg (3/4 per week). UK – Denmark services operated in conjunction with *DFDS Tor Line* who provide all vessels.

1	MAERSK ANGLIA	13017t	00	18.6k	142.5m	12P	-	114T	A	NL
2	MAERSK EXPORTER	13017t	96	18.6k	142.5m	12P	-	114T	A	NL
3	MAERSK FLANDERS	13073t	00	18.6k	142.5m	12P	-	114T	A	NL
4	MAERSK IMPORTER	13017t	96	18.6k	142.5m	12P	-	114T	A	NL
5	UND AKDENIZ	22900t	01	21.6k	193.0m	12P	-	190T	A	TU
6	UND KARADENIZ	22900t	00	21.6k	193.0m	12P	-	190T	A	TU

MAERSK ANGLIA Built by Guangzhou Shipyard International, Guangzhou, China for *Norfolkline*. Entered service as the GUANGZHOU 7130011 (unofficially the 'China II') but renamed shortly afterwards. Operates on the Scheveningen – Felixstowe service.

MAERSK EXPORTER, MAERSK IMPORTER Built by Miho Shipyard, Shimizu, Japan for *Norfolkline*.

Used on the Scheveningen -Felixstowe service.

MAERSK FLANDERS Built by Guangzhou Shipyard International, Guangzhou, China for *Norfolkline*. Used on the Scheveningen - Felixstowe service.

UND AKDENIZ, UND KARADENIZ Built by Flensburger Schiffsbau, Flensburg, Germany for *UND Roro* of Turkey and operated between Istanbul and Trieste. In 2005 sold to *Norfolkline* and chartered back. Likely to be delivered to *Norfolkline* when operations are moved to Vlaardingen. To be reflagged and renamed.

NORSEMERCHANT FERRIES

THE COMPANY, MANAGEMENT AND ADDRESS. See Section 1.

TELEPHONE Administration +44 (0)28 9077 9090, **Reservations *Belfast (Liverpool service)*** +44 (0)870 6099 299, ***(Heysham service)*** +44 (0) +44 (0)870 6099 299, ***Liverpool (Belfast service)*** +44 (0) +44 (0)870 6099 299, ***(Dublin service)*** +44 (0) +44 (0)870 6099 299, ***Heysham*** +44 (0)1524 865050, ***Dublin*** +353 (0)1 819 2955.

FAX Administration *Belfast* +44 (0)28 9077 1286, **Reservations *Belfast (Liverpool service)*** +44 (0)28 9077 5520, ***(Heysham service)*** +44 (0)28 9078 6073, ***Heysham*** +44 (0)1524 865070, ***Liverpool (Belfast service)*** 44 (0)151 906 2718, ***(Dublin service)*** 44 (0)151 906 2718, ***Dublin*** +353 (0)1 819 2941.

INTERNET Email freightbookings@norsemerchant.com **Website** www.norsemerchant.com *(English)*

ROUTES OPERATED Port of Liverpool (Twelve Quays River Terminal, Birkenhead) *(dep: 02.30 Tue-Sat, 10.45 Sun, Tue-Sat, 22.00 Daily*)* – Dublin *(dep: 10.00 Daily*, 14.30 Mon-Sat, 22.00 Sun, 22.45 Mon-Sat)* (3 hrs; (**BRAVE MERCHANT (see Section 1), 2)*; 2 per day), Heysham *(dep: 09.00 Tue-Sat,19 21.00 Daily)* – Dublin *(dep: 09.00 Tue-Sat, 21.00 Daily)* (8 hrs; *(3,7)*; 2 per day), Heysham *(dep: 09.00 Tue-Sat, 19.30 Tue-Sat, 23.30 Daily)* – Belfast *(dep: 07.30 Tue-Sat, 11.30 Tue-Sat, 21.00 Daily)* (7 hrs; *(1,5,6)*; 3 per day). Note: In September 2005 the LINDAROSA will be replaced by the *LIVERPOOL VIKING*, a passenger vessel listed in Section 1).

NorseMerchant Ferries also operate passenger/freight services from Liverpool to Belfast. See Section 1.

1	LEILI	7606t	99	17.0k	122.3m	12P	-	84T	A	ES
2	LINDAROSA	17428t	96	19.7k	183.1m	12P	-	144T	AS	IT
3	MERCHANT BRAVERY	9368t	78	17.0k	133.0m	12P	-	94T	A	JM
4	MERCHANT BRILLIANT	9368t	79	17.0k	133.0m	12P	-	94T	A	JM
5	RIVER LUNE	7765t	83	15.0k	121.4m	12P	-	90T	A	BS
6	SAGA MOON	7746t	84	15.0k	134.8m	12P	-	66T	A	GI
7	VARBOLA	7800t	98	17.0k	122.3m	12P	-	84T	A	ES

LEILI Built by Astilleros de Huelva SA, Huelva, Spain as the LEILI for *Estonian Shipping Company*. Used on Baltic services. In 2002 chartered to *Crowley Maritime* of the USA and renamed the PORT EVERGLADES EXPRESS. In 2004 resumed the name LEILI and chartered to *NorseMerchant Ferries* to operate between Birkenhead and Dublin. In june moves to the Heysham - Belfast route.

LINDAROSA Built by CN Visentini di Visentini Francesco & C, Donanda, Italy for *Levantina Trasporti* of Italy. Chartered to *CoTuNav* of Tunisia for service between Tunisia and Italy. In 2002 chartered to *NorseMerchant Ferries* to operate additional sailings between Liverpool and Belfast. In 2003 transferred to the Birkenhead – Dublin service. In 2005 sold to *Stena RoRo* but continued to be chartered to *NorseMerchant Ferries*.

MERCHANT BRAVERY Built by Nylands Verksted, Oslo, Norway. Launched as the STEVI for *Steineger & Wiik* of Norway and, on delivery, chartered to *Norient Line* of Norway, being renamed the NORWEGIAN CRUSADER. In 1980 chartered to *Ignazio Messina* of Italy for Mediterranean service and renamed the JOLLY GIALLO. In 1982 the charter ended and she was briefly renamed the NORWEGIAN CRUSADER before being purchased by *Ignazio Messina* and resuming the name JOLLY

GIALLO. In 1993 sold to *Merchant Ferries,* renamed the MERCHANT BRAVERY and placed on the Heysham – Warrenpoint (Dublin since 1995) service. In 1999 transferred to *Belfast Freight Ferries'* Heysham – Belfast service. In 2004 moved to the Heysham – Dublin route. In 2004 sold to *Bravery International Shipping Ltd* of Latvia and chartered back to *NorseMerchant Ferries.* Re-registered in Jamaica. Currently operating between Heysham and Dublin.

MERCHANT BRILLIANT Built by Vaagen Verft A/S, Kyrksæterøra, Norway as the NORWEGIAN CHALLENGER *for Steineger & Wiik* of Norway and chartered to *Norient Line* of Norway. In 1982, chartered to *Ignazio Messina* of Italy for Mediterranean service and renamed the JOLLY BRUNO. Later in 1982 she was purchased by *Ignazio Messina.* In 1993 sold to *Merchant Ferries,* renamed the MERCHANT BRILLIANT and placed on the Heysham – Warrenpoint (Dublin since 1995) service. In 1999 transferred to *Belfast Freight Ferries'* Heysham – Belfast service. In 2004 sold to *Brilliant Shipping Ltd* of The Bahamas and chartered back to *NorseMerchant Ferries.* Re-registered in Jamaica. In late March 2005 chartered to *Cobelfret Ferries* for five months.

RIVER LUNE Built by Santierul Naval SA, Galatz, Romania for *Almira Shipping* of Liberia (part of the Norwegian *Balder* group) as the BALDER VIK and initially used on services between Italy and the Middle East. Subsequently she was employed on a number of charters including *North Sea Ferries* and *Norfolk Line.* In 1986 she was acquired by *Navimpex* of Romania, renamed the BAZIAS 7 and initially used on Mediterranean and Black Sea services. In 1987 she was chartered to *Kent Line* for service between Chatham and Zeebrugge. In 1988 she was sold to *Stena Rederi AB* of Sweden and chartered for service between Finland and Germany. In 1989 she was briefly renamed the STENA TOPPER before being further renamed the SALAR. During the ensuing years she undertook a number of charters. In 1993 she briefly resumed the name STENA TOPPER before being chartered to *Belfast Freight Ferries* and renamed the RIVER LUNE. In October 1996 she was sold to *Belfast Freight Ferries.* In 1999 she was transferred to *Merchant Ferries'* Heysham – Dublin service. In 2000 she returned to the Heysham – Belfast route. In January 2003 replaced the VARBOLA on the Heysham – Dublin service. In summer 2003 chartered to *Dart Line,* replacing the VARBOLA which was sub-chartered to *NorseMerchant Ferries.* Currently operating on the Heysham – Belfast service .

SAGA MOON Built by Schlichting Werft GmbH, Travemünde, Germany as the LIDARTINDUR for *Trader Line* of the Faroe Islands for services between Tórshavn and Denmark. In 1986 chartered to *Belfast Freight Ferries* renamed the SAGA MOON. In 1990 she was purchased by *Belfast Freight Ferries.* In 1995 she was lengthened by 18m to increase trailer capacity from 52 to 72 units and trade cars from 25 to 50; the lift was replaced by an internal fixed ramp. In 1998 she was transferred to *Merchant Ferries'* Heysham – Dublin service and in 2001 back to the Heysham – Belfast service. Resumed service between Heysham and Dublin in 2002. In 2004 transferred to the Heysham – Belfast service.

VARBOLA Built by Astilleros de Huelva SA, Huelva, Spain as the VARBOLA for *Estonian Shipping Company.* On completion, chartered to *Dart Line* and placed on the Dartford – Vlissingen route. In 1999 she was renamed the DART 6. At the end of August 1999, the charter was terminated and she was renamed the VARBOLA. She undertook a number of short term charters, including *Merchant Ferries.* In 2000 long-term chartered to *Merchant Ferries* to operate between Heysham and Dublin. In 2003 the charter ended and she was chartered to *Dart Line* to replace the DART 9; she was placed initially on the Dartford – Vlissingen route but later transferred to the Dartford – Dunkerque route. Later sub-chartered to *NorseMerchant Ferries* and placed on the Heysham – Dublin route. In 2004 charter transferred to *NorseMerchant Ferries.*

NORTHLINK ORKNEY AND SHETLAND FERRIES

THE COMPANY, MANAGEMENT, ADDRESS, FAX & INTERNET See Section 1.

TELEPHONE Freight Reservations 0845 6060 449.

ROUTES OPERATED Aberdeen *(dep: 18.00 Sun)* – Kirkwall (Orkney) *(dep: 18.00 Mon)* (8 hrs; *(1,2)*; 1 per week), Aberdeen *(dep: 18.00 Tue-Fri)* – Lerwick (Shetland) *(dep: 18.00 Tue-Fri)* (13 hrs; *(1,2)*; 4 per week).

1	CLARE	5617t	72	17.0k	114.9m	12P	-	62T	A	NO
2	HASCOSAY	6136t	71	17.0k	118.4m	12P	-	50T	A	UK

CLARE Built by Rickmers Werft, Bremerhaven, Germany as the WESERTAL for *Reinecke* of Germany. After delivery she was renamed the MEYER EXPRESS and resumed the name WESERTAL in 1973. She was chartered out to a number of operators including *North Sea Ferries* and *Olau Line*. In 1998 she was briefly renamed the. In 1993 she was sold to Italian Interests and renamed the VINZIA E. She was chartered to *Stena Sealink Line* and operated between Newhaven and Dieppe. In 1994 she was chartered to DFDS subsidiary *Dan-Let Line* (later *DFDS Baltic Line*) and renamed the DANA BALTICA. She operated between Denmark and Lithuania. In 1996 she was renamed the CLARE and again placed on the charter market. 1997 she briefly served on *NorSea Link*, a joint venture between *Scandlines (DSB Rederi)* and *Norse Irish Ferries* between Kristiansand and Eemshaven in the north of the Netherlands. In 1998 she was again chartered to *DFDS* to institute freight only services between Newcastle and IJmuiden. After further brief charters she was, in 1999, briefly chartered to *CargoConnect Transport + Logistics* and placed on a new service between Hull and Hamburg. In 2001, she was chartered to *Smyril Line* to operate between Tórshavn and Hanstholm. In 2002, Lerwick was added to her winter itinerary. From September 2002 Aberdeen was added to a year round roster but dropped at the beginning of 2003. In autumn 2003, the charter was ended. In December 2003 chartered to *NorthLink*.

HASCOSAY Built by Kristiansands M/V A/S, Kristiansand, Norway as the JUNO. In 1979 sold to *Finnfranline* of France, renamed the NORMANDIA and chartered to *Finncarriers* for service between Finland and France. In 1982 chartered to *Sudcargo* and used on services between France and Algeria and the Middle East. In 1986 sold to *Mikkola* of Finland, renamed the MISIDIA and chartered to *Transfennica* for services between Finland and Northern Europe. In 1990 sold to *Kristiania Eiendom* of Norway and renamed the EURO NOR. In 1991 she was chartered to *Commodore Ferries* and renamed the COMMODORE CLIPPER. In 1996 she was replaced by the COMMODORE GOODWILL and renamed the SEA CLIPPER. She was placed on the charter market. In 1998 she was chartered to the *Estonian Shipping Company (ESCO)* and operated between Germany and Estonia; she was renamed the TRANSBALTICA. In 2001 she resumed the name SEA CLIPPER and chartered to *Fjord Line*. In 2002 she was sold to *NorthLink* and renamed the HASCOSAY. She was modified to enable her to accommodate *NorthLink's* cassette system for livestock transport in addition to commercial vehicles. She was chartered to *Caledonian MacBrayne* to operate between Ullapool and Stornoway during summer 2002. In October 2002 entered service with *NorthLink*.

P&O FERRIES

THE COMPANY, MANAGEMENT, ADDRESS, TELEPHONE See Section 1.

INTERNET Website www.poferriesfreight.com www.ponsffreight.com *(English)*

ROUTES OPERATED Dover *(dep: 01.15, 07.40, 11.50, 16.05, 20.25 Daily)* – Calais *(dep: 05.25, 10.40, 14.55, 19.15, 23.35 Daily)* (1hr 20 mins; *(1)*; 5 per day), Hull *(dep: 16.00 Thu (5), 20.00 Tue, Thu (2), 21.00 Daily (passenger ship), 21.00 Wed (3))* – Rotterdam *(dep: 21.00 Daily (passenger ship), 23.30 Mon, Wed, Fri (2))*, Hull *(dep: 19.00 Daily (passenger ship), 21.00 Mon, Wed+, Fri) (3)* – Zeebrugge *(dep: 01.00 Thu (5), 19.00 Daily (passenger ship), 21.00 Tue, Sat, 22.00 Thu (3), 23.00 Wed (2))* (13 hrs (+23 hrs via Rotterdam); 11 per week), Middlesbrough (Teesport) *(dep:17.00 Sat, 21.00 Mon-Wed, Fri)* – Rotterdam (Beneluxhaven, Europoort) *(dep: 19.00 Sat, 23.00 Mon-Tue, Thu-Fri)* (15 hrs; *(4,5)*; 5 per week), Middlesbrough (Teesport) *(dep: 17.00 Sat, 21.00 Mon-Fri)* – Zeebrugge *(dep: 20.30 Mon-Sat)* (15 hrs 30 mins; *(6,7)*; 6 per week).

1	EUROPEAN SEAWAY	22986t	91	21.0k	179.7m	200P	-	120L	BA2	UK
2	LEHOLA	7800t	98	17k	122.3m	12P	-	86T	A	ES
3	NORCAPE	14807t	79	19.4k	151.0m	12P	-	124T	A	NL
4	NORKING	17884t	80	19.0k	170.9m	12P	-	155T	A	FI
5	NORQUEEN	17884t	80	19.0k	170.9m	12P	-	155T	A	FI
6	NORSKY	19992t	99	20.0k	180.0m	12P	-	194T	A	NL
7	NORSTREAM	19992t	99	20.0k	180.0m	12P	-	194T	A	NL

EUROPEAN SEAWAY Built by Schichau Seebeckwerft AG, Bremerhaven, Germany for *P&O European Ferries* for the Dover – Zeebrugge freight service. In 2000 a regular twice daily freight only Dover-Calais service was established, using this vessel, which continued to operate to Zeebrugge at night. In 2001 passengers (not foot or coach passengers) began to be conveyed on the Dover – Zeebrugge service. In 2003 the Zeebrugge service ended and she operated only between Dover and Calais in a freight only mode. In 2004 withdrawn and laid up. In January 2005 returned to the Dover – Calais route.

LEHOLA Built at Huelva, Spain for *Estonian Shipping Company*. Initially used on *ESCO* Baltic services. In 1998 chartered to *Czar Peter Line* to operate between Moerdijk (Netherlands) and Kronstadt (Russia). In 1999 chartered to *Delom* of France to operate between Marseilles and Sete and Tunis. In 2000 she returned to *ESCO*, operating between Kiel and Tallinn. In 2003 purchased by *Scandlines AG* and transferred to subsidiary *Scandlines Estonia AS*. Operated Rostock – Helsinki – Muuga initially and later Rostock – Helsinki. Service finished at the end of 2004 and in 2005 she was chartered to *P&O Ferries* to operate between Hull and Rotterdam.

NORCAPE Built by Mitsui Engineering and Shipbuilding, Tamano, Japan. Launched as the PUMA but, on completion chartered to *B&I Line* and renamed the TIPPERARY for their Dublin – Liverpool service. In 1989 sold to *North Sea Ferries*, renamed the NORCAPE and introduced onto the Ipswich – Rotterdam service. In 1995 that service ceased and she was moved to the Hull – Zeebrugge freight service. She retains Dutch crew and registry.

NORKING, NORQUEEN Built by Rauma-Reopla Oy, Rauma, Finland as the BORE KING and the BORE QUEEN for *Bore Line* of Finland for Baltic services. In 1991 chartered to *North Sea Ferries* for their Teesport – Zeebrugge service and renamed the NORKING and NORQUEEN respectively. During winter 1995/96 they were lengthened by 28.8 metres and re-engined. In 1999 transferred to the Teesport – Rotterdam service.

NORSKY, NORSTREAM Built by Aker Finnyards, Rauma, Finland for *Bore Line* of Finland and chartered to *P&O North Sea Ferries*. They operate on the Teesport – Zeebrugge service.

P&O IRISH SEA

THE COMPANY AND ADDRESS See Section 1.

MANAGEMENT Chairman Russ Peters, **Managing Director** J H Kearsley, **Commercial Manager** Philip Simpson.

TELEPHONE Administration +44 (0)1253 615700, **Reservations *UK*** 0870 6000 868, ***Irish Republic*** +353 (0)1 855 0522.

FAX Administration & Reservations *Cairnryan* +44 (0)1581 200282, ***Larne*** +44 (0)28 2827 2477, ***Fleetwood*** +44 (0)1253 615740.

INTERNET Website www.poisfreight.com *(English)*

ROUTES OPERATED Troon *(dep: 02.30 Mon-Sat, 12.00 Sun)* – Larne *(dep: 10.00 Sat, 19.00 Sun-Fri)* (4 hrs 30 mins; *(2)*; 1 per day).

1	EUROPEAN MARINER	5897t	77	15.0k	116.3m	12P	-	62T	A	BS

EUROPEAN MARINER Built at Bremerhaven, Germany as the SALAHALA and chartered to *Gilnavi* of Italy for Mediterranean services. In 1990 she was purchased by *Cenargo* and chartered to *Merchant Ferries* who renamed her the MERCHANT VALIANT. She was used on their Fleetwood – Warrenpoint

Begonia *(FotoFlite)*

Stena Partner *(Andrew Cooke)*

service until 1993 when she was chartered to *Pandoro* and placed on their Ardrossan – Larne service. Purchased by *P&O* in 1995 and renamed the LION. In early 1998 renamed the EUROPEAN HIGHLANDER. In July 2001, the service moved to Troon and she was renamed the EUROPEAN MARINER. In 2002 replaced by the EUROPEAN NAVIGATOR. After a brief charter to *Seatruck Ferries* she went on a two month charter to *Color Line*, operating between Kristiansand and Hirtshals. On return she went on a short charter to *Commodore Ferries* and in late September was transferred to the Larne – Troon route replacing the EUROPEAN NAVIGATOR. Later she was chartered to *Norse Island Ferries*; she returned to *P&O Irish Sea* in late 2002.

SCA TRANSFOREST

THE COMPANY *SCA Transforest* is a Swedish company.

MANAGEMENT Managing Director (UK) Bo Frölander.

ADDRESS Interforest Terminal London Ltd, 44 Berth, Tilbury Dock, Essex RM18 7HR.

TELEPHONE Administration & Reservations *Sweden* +46 (0)60-19 35 00, ***UK*** +44 (0)1375 48 85 00.

FAX Administration & Reservations *Sweden* +46 (0)60-19 35 74 ***UK*** +44 (0)1375 48 85 03.

INTERNET Email *Sweden* info@transforest.sca.com ***UK*** bo.frolander@sca.com

Website www.transforest.sca.se *(English)*

ROUTE OPERATED Umeå *(dep: 11.00 Mon, 16.00 Thu)* – Husum *(dep: 20.00 Mon, 22.00 Thu)* – Sundsvall *(dep: 12.00 Tue, 12.00 Fri)* – Iggesund *(dep: 19.00 Tue, 19.00 Fri)* – Tilbury *(arr: 11.00 Tue, 13.00 Sat, dep: 16.00 Tue, 18.00 Sat)* – Rotterdam (Eemhaven) *(dep: 13.00 Sun, 12.00 Wed)* – Helsingborg *(arr: 07.00 Fri)* (8/9 day round trip; *(1,2,3)*; 2 per week).

1	OBBOLA	20171t	96	16.0k	170.6m	0P	-	-	A	SW
2	ORTVIKEN	20171t	97	16.0k	170.4m	0P	-	-	A	SW
3	ÖSTRAND	20171t	96	16.0k	170.6m	0P	-	-	A	SW

OBBOLA, ORTVIKEN, ÖSTRAND Built by Astilleros Españoles, Seville, Spain for *Gorthon Lines* and chartered to *SCA Transforest*. They are designed for the handling of forest products in non-wheeled 'cassettes' but can also accommodate ro-ro trailers; however no trailer capacity is quoted. The ORTVIKEN was lengthened during autumn 2000 and the OBBOLA and ÖSTRAND were lengthened during 2001.

SEA-CARGO

THE COMPANY *Sea-Cargo AS* of Norway is a joint venture between *Nor-Cargo AS* (a Norwegian company jointly owned by *Ofotens og Vesteraalen Dampskipsselskab, Det Stavangerske Dampskipsselskab* and *Troms Fylkes Dampskipsselskab*) and *SeaTrans DS* of Norway.

MANAGEMENT *Sea-Cargo UK Ltd* Managing Director Barry Jenks.

ADDRESS *Norway* Sea-Cargo AS, PO Box 353, Nesttun, 5853 Bergen, Norway, ***Immingham*** Sea-Cargo UK, West Riverside Road, Immingham Dock, Immingham DN40 2NT, ***Aberdeen*** Nor-Cargo Ltd, Matthews Quay, Aberdeen Harbour, Aberdeen, AB11 5PG.

TELEPHONE Administration & Bookings *Bergen* +47 55 10 84 84, ***Immingham*** +44 (0)1469 577119, ***Aberdeen*** +44 (0)1224 596481.

FAX Administration & Reservations *Bergen* +47 55 91 22 33, ***Immingham*** 44 (0)1469 577708, ***Aberdeen*** +44 (0)1224 582360.

INTERNET Email mail@sea-cargo.no **Website** www.sea-cargo.no *(English, Norwegian)*

ROUTES OPERATED *Circuit 1* Bergen *(dep: 17.00 Thu)* – *Knarrevik *(arr: 18.00 Thu, dep: 21.00 Thu)* – Odda *(arr: 06.00 Fri, dep: 16.00 Fri)* – Haugesund *(arr: 00.01 Sat, dep: 01.00 Sat)* – Tanager *(arr: 02.00 Sat, dep: 14.00 Sat)* – Immingham *(arr: 01.00 Mon , dep: 21.00 Tue)* – Tananger *(arr:*

03.00 Thu, dep: 05.00 Thu) – Bergen *(arr: 13.00 Thu)* (1 week; *(1)*; weekly), ***Circuit 2*** Bergen *(dep: 14.00 Tue)* – Husnes *(arr: 18.00 Tue, dep: 22.00 Tue)* – Kamøy *(dep: Wed)* – Tanager *(arr: 07.00 Wed, dep: 18.99 Wed)* – Tanager (arr: 21.00 Wed, dep: 22.00 Wed) – Amsterdam *(arr 07.00 Fri, dep: 21.00 Fri)* – Immingham *(arr: 15.00 Sat, dep: 21.00 Sat)* – Tanager *(arr: 08.30 Mon, dep: 09.00 Mon)* – Bergen *(arr: 15.00 Mon)* (6 days; *(4)*; weekly), ***Circuit 3*** Bergen *(dep: 13.00 Tue)* – Florø *(arr: 20.30 Tue, dep: 21.00 Tue)* –Aalesund *(arr: 02.00 Wed, dep: 05.00 Wed)* – Trondheim *(arr: 15.00 Wed, dep: 20.00 Wed eve)* – Kristiansund *(arr: 02.00 Thu, dep: 05.00 Thu)* – Molde *(arr: 08.00 Thu, dep: 10.00 Thu)* – Aalesund *(arr: 13.00 Thu, dep: 16.00 Thu)* – Bergen *(arr: 12.00 Fri, dep: 18.00 Fri)* – Tanager *(arr: 01.00 Sat, dep: 06.00 Sat)* – Aberdeen *(arr: 01.00 Sun, dep: 06.00 Sun)* – Tanager *(arr: 03.00 Mon, dep: 06.00 Mon)* – Haugesund *(arr: 08.30 Mon, dep: 09.00 Mon)* – Bergen *(arr: 19.00 Mon)* (6 days; *(2)*; weekly), ***Circuit 4*** Bergen *(dep: 05.00 Fri)* – *Knarrevik *(arr: 07.00 Fri, dep: 16.00 Fri)* – Bergen *(arr: 17.00 Fri, dep: 19.00 Fri)* – Tanager *(arr: 05.00 Sat, dep: 10.00 Sat)* – Amsterdam *(arr: 18.00 Sat, dep: 22.00 Sat)* – *Rotterdam *(arr: 06.00 Mon, dep: 12.00 Mon)* – *Moerdijk *(arr: 16.00 Mon, dep: 22.00 Mon)* – Amsterdam *(arr: 08.00 Tue, dep: 23.00 Tue)* – Tanager *(arr: 07.00 Thu, dep: 10.00 Thu)* – Bergen *(arr: 18.00 Thu)* (7 days; *(3)*; weekly). *Every second voyage.

1	COMETA	4610t	81	16.0k	102.2m	0P	-	26T	AS	NO
2	SC ABERDEEN	4234t	79	15.5k	109.0m	0P	-	24T	AS	BS
3	SC NORRLAND	5562	90	14.5k	107.5m	0P	-	24T	AS	AN
4	TRANS CARRIER	8476t	93	14.5k	125.2m	0P	-	78T	A	BS

COMETA Built by Fosen Mekaniske Verksted, Rissa, Norway for *Nor-Cargo*.

SC ABERDEEN Built by Fosen Mekaniske Verksted, Rissa, Norway for *Nor-Cargo*. Launched as the ERIC JARL but renamed the ASTREA before entering service. In 1986 she sank and, after raising and refitting she was, in 1992, renamed the TUNGENES. In 2001 she was renamed the SC ABERDEEN.

SC NORRLAND Built by J J Sietas KG, Hamburg, Germany as the MARTHA RUSS *Ernst Russ* of Germany and chartered to *Transfennica*. In 1999 sold to *Speedster Shipping Company* chartered to *Baltmed Shipping* of Sweden, renamed the MED STAR and operated between Sweden and Spain. In 2002 renamed the NORRLAND. In 2005 chartered to *Sea-Cargo* and renamed the SC NORRLAND.

TRANS CARRIER Built by Nrodogradiliste Kraljevica, Kraljevica, Croatia as the KORSNÄS LINK for *SeaLink AB* of Sweden and due to be time chartered to *Korsnäs AB*, a Swedish forest products company. However, due to the war in Croatia, delivery was seriously delayed and she was offered for sale. In 1994 sold to the *Swan Group* and renamed the SWAN HUNTER. She was placed on the charter market. In 1997 she was chartered to *Euroseabridge* and renamed the PARCHIM. In 1999 the charter ended and she resumed the name SWAN HUNTER. In 1999 she was sold to *SeaTrans* and renamed the TRANS CARRIER.

SEAFRANCE

THE COMPANY, MANAGEMENT & ADDRESS See Section 1.

TELEPHONE Reservations +44 (0)1304 203030.

FAX Reservations +44 (0)1304 212726

INTERNET Email freightdover@wanadoo.fr **Website** www.seafrancefreight.com *(English, French)*

ROUTE OPERATED Calais *(dep: 02.15 Mon-Sat, 09.15 Mon-Sat, 13.30, 17.45, 22.00 Sun-Fri)* – Dover *(dep: 06.00 Mon-Sat, 10.15 Mon-Sat, 14.30, 18.45, 23.00 Sun-Fri)* (1 hr 30 mins; *(1)*; 5 per day).

1	SEAFRANCE NORD PAS-DE-CALAIS	13727t	87	21.5k	160.1m	80P	-	102L	BA2	FR

SEAFRANCE NORD PAS-DE-CALAIS Built by Chantiers du Nord et de la Mediterranee, Dunkerque, France as the NORD PAS-DE-CALAIS at Dunkerque, France for *SNCF* for the Dunkerque (Ouest) – Dover train ferry service. Before being used on this service (which required the construction of a new berth at Dover (Western Docks)) in May 1988, she operated road freight services from Calais to Dover Eastern Docks. The train ferry service continued to operate following the opening of the Channel Tunnel in 1994, to convey road vehicles and dangerous loads which were banned from the tunnel. However, it ceased in December 1995 and, after a refit, in February 1996 she was renamed the SEAFRANCE NORD PAS-DE-CALAIS and switched to the Calais – Dover service, primarily for road freight vehicles and drivers but also advertised as carrying up to 50 car passengers. Since the entry into service of a third multi-purpose ferry, she has operated on a freight-only basis.

SEATRUCK FERRIES

THE COMPANY *Seatruck Ferries Ltd* is a British private sector company. It is part of the *Clipper Group*.

MANAGEMENT Managing Director Kevin Hobbs, **Sales Director** Alistair Eagles.

ADDRESS *Warrenpoint (HQ)* Seatruck House, The Ferry Terminal, Warrenpoint, County Down BT34 3JR. ***Heysham*** North Quay, Heysham Port, Heysham, Morecambe, Lancs LA3 2UL.

TELEPHONE Administration +44 (0)28 4175 4411, **Reservations *Warrenpoint*** +44 (0)28 4175 4400, ***Heysham*** +44 (0)1524 853512.

FAX Administration +44 (0)28 4175 4545, **Reservations *Warrenpoint*** +44 (0)28 4177 3737, ***Heysham*** +44 (0)1524 853549.

INTERNET Email alistair@seatruck-ferries.co.uk **Website** www.seatruckferries.com *(English)*

ROUTE OPERATED Heysham *(dep: 08.00 Tue-Sat, 19.00 Tue-Sat, 23.00 Daily)* – Warrenpoint *(dep: 08.00 Tue-Sat, 11.00 Tue-Sat, 17.00 Sun, 20.00 Mon, Wed-Sat, 21.00 Tue)* (8 hrs; *(1,2,3)*; 3 per day).

1	CHALLENGE	7800t	98	17.0k	122.3m	12P	-	84T	A	ES
2	MOONDANCE	5881t	78	15.0k	116.3m	12P	-	62T	A	BS
3	RIVERDANCE	6041t	77	15.0k	116.3m	12P	-	62T	A	BS

CHALLENGE Built at Huelva A, Spain for *Estonian Shipping Company* as the LEMBITU. On completion chartered to *P&O European Ferries (Irish Sea)* and placed on their Liverpool Dublin route. In Autumn 1998 she was chartered to *Dart Line* and placed on the Dartford – Vlissingen route. In 1999 she was renamed the DART 7. In Autumn 1999 the charter was ended and she was chartered to *Cetma* of France, resumed the name LEMBITU and used on services between Marseilles and Tunis. In 2000 she was chartered to *P&O European Ferries (Irish Sea)* and renamed the CELTIC SUN; she operated between Liverpool and Dublin. In 2001 the chartered ended; she then reverted to the name LEMBITU and was chartered to *NorseMerchant Ferries* and placed on the Heysham – Dublin service. In late 2001 charter ended and returned to *ESCO* service in the Baltic. In 2003 chartered to *Scandlines AG* and placed on their Rostock – Helsinki – Muuga service. This service finished in December 2004 and she was chartered to *Channel Freight Ferries* in January 2005. In March 2005 chartered to *NorseMerchant Ferries* again and operated between Heysham and Belfast and Heysham and Dublin. Purchased by *Elimira Shipping* of Greece and renamed CHALLENGE. In June 2005 chartered to *Seatruck Ferries* for two years.

MOONDANCE Built by Rickmers Werft, Bremerhaven, Germany as the EMADALA for *Emadala Shipping* and chartered to *Gilnavi Line* of Italy for Mediterranean service. In 1987 she was purchased by *Gilnavi Line*. In 1990 sold to *Cenargo* of Great Britain and chartered to *Merchant Ferries* for their Heysham – Warrenpoint service and renamed the MERCHANT VICTOR. She was withdrawn from that service in 1993 and was chartered out to a number of operators. In 1997 she was chartered to *Seatruck Ferries* and renamed the MOONDANCE. In 1998 she was purchased by *Seatruck Ferries*. Following collapse of ramp at Warrenpoint in January 2001, she briefly operated between Heysham and Larne.

RIVERDANCE Built by Rickmers Werft, Bremerhaven, as the MASHALA for *Mashala Shipping* and chartered to *Gilnavi* of Italy for Mediterranean services. After a long period out of service in the mid-nineteen eighties, in 1987 she was sold, renamed the HALLA and chartered for Caribbean service. In 1988 she was renamed the TIKAL. In 1989 she was sold to *Schiaffino Line* of France, renamed the SCHIAFFINO and put into service between Ramsgate and Ostend. In 1990 the company was taken over by *Sally Ferries* and in 1991 she was chartered to *Belfast Freight Ferries*. In 1993 she was renamed the SALLY EUROBRIDGE. In January 1994, she was chartered to *North Sea Ferries* to operate between Hull and Zeebrugge and renamed the EUROBRIDGE. In summer 1994 she returned to *Sally Ferries*, resumed the name SALLY EUROBRIDGE and became the second vessel on the Ramsgate – Vlissingen service; in the autumn the British terminal was switched to Dartford. In 1995 she was chartered to *Norfolk Line*, renamed the EUROBRIDGE and also sold by *Sally Ferries*. In 1996 she was chartered to *Seatruck Ferries* and renamed the RIVERDANCE. In 1997 she was purchased by *Seatruck Ferries*. Following collapse of ramp at Warrenpoint in January 2001, she briefly operated between Heysham and Larne.

Under Construction

4	NEWBUILDING 1	-	07	22.0k	140.0m	12P	-	120T	A	-
5	NEWBUILDING 2	-	07	22.0k	140.0m	12P	-	120T	A	-

NEWBUILDING 1, NEWBUILDING 2 Under construction by Astilleros de Huelva SA, Huelva, Spain for *Seatruck Ferries*. To replace the MOONDANCE and RIVERDANCE.

STENA LINE

THE COMPANY, MANAGEMENT, ADDRESS, TELEPHONE AND INTERNET See Section 1.

ROUTES OPERATED Harwich *(dep: 05.00 (Tue-Sat), 11.00 (Mon-Sat), 22.45 (Sun-Fri)* – Rotterdam *(dep: 11.30 (Mon-Sat), 19.00 (Mon-Fri), 23.45 (Sun-Fri))* (7 hrs 45 mins; *(2,7,8)*; 3 per day), Immingham (Killingholme) *(dep: 19.15)* – Hoek van Holland *(dep: 19.00)* (13 hrs; *(5,6)*; 1 per day), Fleetwood *(dep: 03.00 Mon-Sat, 10.00 Daily, 22.00 Daily)* – Larne *(dep: 10.00 Daily, 16.00 Sun, Mon-Fri, 22.00 Daily)* (7 hrs; *(1,3,4)*; 3 per day). **Note** A limited number of cars and car passengers is carried on the Fleetwood – Larne service.

1	STENA LEADER	12879t	75	17.0k	157.2m	50P	-	114T	A	BD
2	STENA PARTNER	21162t	77	16.5k	184.6m	166P	-	180T	A2	UK
3	STENA PIONEER	14387t	75	17.7k	141.8m	76P	-	114T	A	BD
4	STENA SEAFARER	10957t	75	18.0k	141.8m	50P	-	80T	A	BD
5	STENA SEARIDER	21019t	69	17.0k	178.9m	120P	-	198T	AS2	IM
6	STENA SEATRADER	17991t	73	17.5k	181.6m	221P	-	174T	AS2	UK
7	STENA TRANSFER	21162t	77	16.5k	184.6m	166P	-	180T	A2	UK
8	STENA TRANSPORTER	16776t	78	17.0k	151.0m	74P	-	122T	A2	UK

STENA LEADER Built by J J Sietas KG Schiffswerke, Hamburg, Germany for *Stena AB* as the BUFFALO and due to be chartered to *P&O* for *Pandoro* Irish Sea services. Before completion she was purchased by *P&O*. In 1989 she was lengthened by 12.5m and in 1998 she was further lengthened by 15m and renamed the EUROPEAN LEADER. In 2004 sold to *Stena Line* and renamed the STENA LEADER. Used on the Fleetwood – Larne service.

STENA PARTNER Built by Hyundai Shipbuilders & Heavy Industries, Ulsan, South Korea for *Stena Rederi* as the ALPHA ENTERPRISE and chartered to *Aghiris Navigation* of Cyprus. In 1979 she was renamed the SYRIA and chartered to *Hellas Ferries* for services between Greece and Syria. In 1981 she was lengthened by 33.6m. In 1982 she was chartered to *European Ferries* and used on freight services between Felixstowe and Rotterdam. In 1983 she was renamed the STENA TRANSPORTER and in 1986 the CERDIC FERRY. In 1992 she was renamed the EUROPEAN FREEWAY and, in 1994, purchased by *P&O European Ferries*. In 2002 sold to *Stena Line* and renamed the FREEWAY. She initially operated between Felixstowe and Rotterdam and later Harwich and Rotterdam. In early 2003 she was renamed the STENA PARTNER.

STENA PIONEER Built by J J Sietas KG Schiffswerke, Hamburg, Germany for *Stena AB* as the BISON and due to be chartered to *P&O* for *Pandoro* Irish Sea services. Before completion she was purchased by *P&O*. Between 1989 and 1993 she was operated by *B&I Line* of Ireland on a joint service with *Pandoro* between Dublin and Liverpool. An additional deck was added in 1995. In late 1997 she was renamed the EUROPEAN PIONEER. In 2004 sold to *Stena Line* and renamed the STENA PIONEER. Used on the Fleetwood – Larne service.

STENA SEAFARER Built by J J Sietas KG Schiffswerke, Hamburg, Germany. Ordered by *Stena AB* as the UNION TRADER but completed as the UNION MELBOURNE for the *Northern Coasters Ltd* of the UK and lengthened before entering service. Chartered to the *Union Steamship Company* of New Zealand and used on services to Australia. In 1980 she was sold to another *P&O* subsidiary and renamed the PUMA. In early 1998 she was renamed the EUROPEAN SEAFARER. Used on the Fleetwood – Larne service in recent years, in 2001 she was transferred to the Rosslare – Cherbourg service. In 2002 replaced by the EUROPEAN DIPLOMAT and returned to the Fleetwood – Larne service. In 2004 sold to *Stena Line* and renamed the STENA SEAFARER.

STENA SEARIDER Built by Oy Wärtsilä Ab, Helsinki, Finland as the FINNCARRIER for *Finnlines* of Finland for service between Finland, Denmark and Germany. In 1975 renamed the POLARIS. In 1984 sold to *Rederi AB Nordö* of Sweden to operate between Malmö (Sweden) and Travemünde (Germany) and renamed the SCANDINAVIA. In 1987 she was rebuilt to increase capacity from 122 trailers to 200. In 1989 the name of the company was changed to *Nordö Link* and she was renamed the SCANDINAVIA LINK. In 1990 she was sold to *Stena Line*, renamed the STENA SEARIDER and used on their Gothenburg (Sweden) – Travemünde service. In 1991 she was chartered out for service in the Caribbean and renamed the SEARIDER. In 1992 she was chartered to *Norse Irish Ferries* and renamed the NORSE MERSEY. In 1995 she was replaced by a new vessel of the same name and returned to *Stena Line*, resumed the name STENA SEARIDER and resumed operating between Gothenburg and Travemünde and Gothenburg and Kiel. In May 1997, she was transferred to the Harwich – Hoek van Holland service. In autumn 2000 she inaugurated (with the chartered ROSEBAY – see the TRANSLANDIA, *Eckerö Line*) a new service from Hoek van Holland to Killingholme (near Immingham).

STENA SEATRADER Built by A/S Nakskov Skipsværft, Nakskov, Denmark as the SVEALAND for *Lion Ferry AB* of Sweden and chartered to *Statens Järnvägar (Swedish State Railways)* for the train ferry service between Trelleborg (Sweden) and Sassnitz (Germany (DDR)). The charter ceased in 1980 and in 1982 she was sold to *Rederi AB Nordö* of Sweden. She was lengthened by 33.7 metres, renamed the SVEALAND AV MALMÖ and used on their lorry/rail wagon service between Malmö and Travemünde. In 1986 she was rebuilt with a higher superstructure and in 1987 she was renamed the SVEA LINK, the service being renamed *Nordö Link*. In 1990 she was sold to *Stena Line*, renamed the STENA SEATRADER and introduced onto the Hoek van Holland – Harwich service. In spring 2001 she replaced the chartered ROSEBAY (see TRANSLANDIA, *Eckerö Line*) on the Hoek van Holland – Killingholme service.

STENA TRANSFER Built by Hyundai Shipbuilders & Heavy Industries, Ulsan, South Korea. Launched as the STENA RUNNER by *Stena Rederi* of Sweden. On completion, renamed the ALPHA PROGRESS and chartered to *Aghiris Navigation* of Greece. In 1979 renamed the HELLAS and operated by *Soutos-Hellas Ferry Services* on services between Greece and Syria. In 1982 she was lengthened by 33.6m. In 1982 she was chartered to *European Ferries* and used on freight services between Felixstowe and Rotterdam. The following year she was returned to *Hellas Ferries*. In 1985 she returned to *European Ferries* and the Rotterdam service. In 1986 she was renamed the DORIC FERRY. In 1992 she was renamed the EUROPEAN TIDEWAY and, in 1994, purchased by *P&O European Ferries*. In 2001 replaced by the NORBANK and laid up. In 2002 returned to the Felixstowe – Rotterdam route when the NORBANK was transferred to *P&O Irish Sea*. In 2002 sold to *Stena Line* and renamed the IDEWAY. She initially operated between Felixstowe and Rotterdam and later Harwich and Rotterdam. Later in 2002 she was renamed the STENA TRANSFER.

STENA TRANSPORTER Built by Hyundai Shipbuilders & Heavy Industries, Ulsan, South Korea as the MERZARIO ESPANIA for *Stena Rederi* of Sweden and immediately chartered to *Merzario Line* for their service between Italy and Saudi Arabia. In the same year she was renamed the MERZARIO HISPANIA. In 1979 she was chartered to *European Ferries* for their ro-ro freight service between Felixstowe and Rotterdam and renamed the NORDIC FERRY. In 1982 she served in the Falkland Islands Task Force. In 1986 she was modified to carry 688 passengers and, with sister vessel the BALTIC FERRY (now the EUROPEAN DIPLOMAT), took over the Felixstowe – Zeebrugge passenger service. In 1992 she was renamed the PRIDE OF FLANDERS. In 1994, purchased by *P&O European Ferries*. In 1995 the Felixstowe – Zeebrugge passenger service ceased, her additional passenger accommodation was removed, passenger capacity was reduced and she was transferred to the Felixstowe – Rotterdam freight service. In 2002 sold to *Stena Line* and renamed the FLANDERS. She initially operated between Felixstowe and Rotterdam and later Harwich and Rotterdam. In autumn 2002 chartered to *Scandlines AB* of Sweden to operate between Travemünde and Trelleborg whilst the SVEALAND was undergoing a major rebuild. Later in 2002, after a refit, she returned to the Harwich – Rotterdam route and was renamed the STENA TRANSPORTER. She has also operated between Rotterdam and Immingham (HST, Killingholme).

Under Construction

9	STENA FORWARDER	25900t	06	22.2k	212.0m	200P	-	220T	A	UK
10	STENA TRAVELLER	25900t	06	22.2k	212.0m	200P	-	220T	A	UK

STENA FORWARDER, STENA TRAVELLER Under construction by Fosen Mekaniske Verksted, Rissa, Norway (hulls built at St Petersburg, Russia) for the Hoek van Holland – Immingham (Killingholme) service.

TRANSEUROPA FERRIES

THE COMPANY *TransEuropa Ferries NV* is a Belgian subsidiary of *TransEuropa Shipping Lines*, a Slovenian private sector company. Channel operations started in 1997, in conjunction with *Sally Ferries*, replacing them on November 1998. The company traded as *TransEuropa Shipping Lines (TSL)* until 2000. Note that all the owning companies listed here are associated companies of *TSL*.

MANAGEMENT *TransEuropa Shipping* Managing Director Stergulc Rihard, ***TransEuropa Ferries NV*, General Manager Belgium & UK** Mr Dominique Penel, **Sales Manger, Europe** Mr Peter Sys.

ADDRESS *TSL Slovenia* Vojkovo nabrezje 38, 6000 Koper, Slovenia, ***TEF UK*** Ferry Terminal, Ramsgate New Port, RAMSGATE, Kent CT11 8RP ***TEF Belgium*** Slijkensesteenweg 2, 8400 Ostend, Belgium.

TELEPHONE Admin *TSL Slovenia* +386 (0)5 664 17 77, ***TEF UK*** +44 (0)1843 853833, ***TEF Belgium*** +32 (0)59 34 02 40. **Reservations *TEF UK*** +44 (0)1843 853523, ***TEF Belgium*** +32 (0)59 340250.

FAX *TSL* Admin *Slovenia* +386 (0)5 639 50 36, ***TEF Belgium*** +32 (0)59 34 02 41, **Admin & Sales *TEF UK*** +44 (0)1843 853668, **Sales *TEF Belgium*** +32 (0)59 34 02 51.

INTERNET Website www.transeuropaferries.co.uk *(English)*

ROUTE OPERATED Ramsgate *(dep: 01.00 Sat, 02.00 Mon-Fri, 03.00 Sat, 03.30 Tue-Fri, 07.00 Daily, 09.30 Mon, 12.00 Sun, 13.30 Mon-Fri, 15.30 Sun, 16.30 Mon-Fri, 18.30 Daily, 20.30 Sun-Fri, 22.30 Daily, 23.45 Sun, 23.59 Mon-Thu)* – Ostend (Belgium) *(dep: 01.00 Daily, 03.00 Tue-Thu, 05.30 Mon, 07.00 Daily, 11.00 Sun-Fri, 13.30 Daily, 16.00 Sun-Fri, 18.00 Daily, 20.00 Sun-Thu, 21.30 Sun-Thu, 22.30 Fri, 23.00 Mon-Thu, 23.30 Sun)* (4 hrs; *(1,2,3 plus passenger vessels EUROVOYAGER, LARKSPUR and OLEANDER)*; up to 10 per day).

1	BEGONIA	8023t	76	18.5k	118.1m	105P	-	52L	BA	SV
2	GARDENIA	8097t	78	18.4k	118.1m	105P	-	52L	BA2	CY
3	PRIMROSE	12046t	75	22.0k	118.4m	1250P	348C	45L	BA2	CY

BEGONIA Built by Schichau-Unterweser AG, Bremerhaven, Germany as the EUROPEAN CLEARWAY for *European Ferries* ro-ro freight services. She was used on freight services between Dover and Calais and Dover and Zeebrugge. In 1992 she was moved to the Portsmouth – Le Havre route. In 1993 she was transferred to *Pandoro* to inaugurate a new Cherbourg – Rosslare service. In 1996 she was renamed the PANTHER. In early 1998 she was renamed the EUROPEAN PATHFINDER. In 2001 she was moved to the Cairnryan – Larne service to replace the EUROPEAN TRADER. In 2002 sold *ERATO Shipping* and renamed the REGINA I. Before delivery, resold to *Abbey Trading SA Trust Co*, delivered to *TEF* and renamed the BEGONIA. In 2004 began operating for *TransEuropa Ferries*.

GARDENIA Built by Schichau-Unterweser AG, Bremerhaven, Germany as the EUROPEAN ENTERPRISE for *European Ferries*. In 1988 she was renamed the EUROPEAN ENDEAVOUR. She was used on freight services between Dover and Calais and Dover and Zeebrugge. If space was available, a small number of passengers was sometimes conveyed on the Zeebrugge service, although the sailings were not advertised for passengers. This ceased with the withdrawal of passenger services on this route at the end of 1991. During the summer period she provided additional freight capacity on the Dover – Calais service and has also served on other routes. In autumn 1995 she was transferred to the Cairnryan – Larne service. In 1998 accommodation was raised to provide extra freight capacity. In March 1999 began also operating from Larne to Ardrossan but this ceased later in the year. Withdrawn from service in July 2002 and sold to *Odyssy Maritime Co Ltd* and renamed the GARDENIA. In 2003 she began operating for *TEF* between Ramsgate and Ostend.

PRIMROSE Built by NV Cockerill Yards Hoboken, Hoboken, Belgium as the PRINCESSE MARIE-CHRISTINE for *Regie voor Maritiem Transport* of Belgium for the Ostend – Dover service. During 1985 she had an extra vehicle deck added, increasing vehicle capacity. Passenger capacity was increased by 200 by the conversion of an upper deck 'garage' into passenger accommodation. In January 1994 the British port became Ramsgate. In 1994 chartered briefly to *Sally Ferries* and operated between Ramsgate and Dunkerque. Since then a spare vessel and withdrawn in early 1997. In 1998 sold to *Dianthus Maritime Co Ltd* of the UK and renamed the PRIMROSE. In 1999 she began operating for *TSL* between Ramsgate and Ostend after a major refit at Dunkerque.

TRANSFENNICA

THE COMPANY *Transfennica Ltd* is a Finnish private sector company.

MANAGEMENT President Rolf G W Eriksson, **Director (UK)** Jim Deeprose, **Operations Manager (UK)** Andrew Prior.

ADDRESS *Finland* Eteläranta 12, 00130 Helsinki, Finland, ***UK*** Finland House, 47 Berth, Tilbury Freeport, Tilbury, Essex RM18 7EH.

TELEPHONE Administration & Reservations *Finland* +358 (0)9 13262, ***UK*** +44 (0)1375 363 900.

FAX Administration & Reservations *Finland* +358 (0)9 652377, ***UK*** +44 (0)1375 840 888.

INTERNET Email *Finland* info@transfennica.com ***UK*** info.uk@transfennica.com

Website www.transfennica.com *(English)*

ROUTES OPERATED *Circuit 1* – Hamina *(dep: 17.00 Thu)* – Hanko *(arr: 07.00 Fri, dep 14.00 Fri)* – Tilbury *(arr: 06.00 Mon, dep: 17.00 Mon)* – Hamina *(arr: 06.00 Thu)*; ***Circuit 2*** Hamina *(dep: 18.00 Mon)* – Tilbury *(06.00 Fri, 17.00 Fri)* – Hamina *(arr: 10.00 Mon)*; ***Circuit 3*** – Rauma *(dep: 18.00 Fri)* – Antwerp *(arr: 08.00 Mon, dep: 10.00 Mon)* – Tilbury (*arr: 08.00 Tue, dep: 16.00 Tue)* – Rauma *(arr: 07.00 Fri)*; *(1,2,3,4)*. Note: 'dep' times are closure times for freight. Ship will actually leave a little later. Ships will often arrive at Tilbury before scheduled time – sometimes the night before if coming from Finland.

1	CAROLINE RUSS	10471t	99	21.0k	153.5m	12P	-	120T	A2	AT
2	ELISABETH RUSS	10471t	99	21.0k	153.5m	12P	-	120T	A2	AT
3	PAULINE RUSS	10471t	99	21.0k	153.5m	12P	-	120T	A2	AT
4	SEAGARD	10471t	99	21.0k	153.5m	12P	-	134T	A2	FI

CAROLINE RUSS, ELISABETH RUSS, PAULINE RUSS Built by J J Sietas KG, Hamburg, Germany for *Ernst Russ* of Germany and chartered to *Transfennica*.

SEAGARD Built by J J Sietas KG, Hamburg, Germany for *Bror Hussel* of Finland and chartered to *Transfennica*.

BRITISH ISLES
Scrabster
Stornoway
Ullapool
Aberdeen
Castlebay
Mallaig
Oban
Rosyth
Gourock
Largs
Troon
Cairnryan
Stranraer
Newcastle
Middlesbrough
Arranmore
Larne
Belfast
Douglas
Heysham
Fleetwood
Portaferry
Hull
Killingholme
Immingham
Warrenpoint
Holyhead
Liverpool
Dublin
Dun Laoghaire
Tarbert
Rosslare
Fishguard
Castletownbere
Passage East
Cork
Pembroke Dock
Felixstowe
Harwich
Purfleet
Sheerness
Dartford
Ramsgate
Dover
Southampton
Portsmouth
Newhaven
Weymouth
Poole
Fowey
Dartmouth
Penzance

section 4 chain, cable etc ferries

gb & ireland

In addition to the ferries listed above, there are a number of short chain ferries, cable ferries and ferries operated by unpowered floats.

BOURNEMOUTH-SWANAGE MOTOR ROAD AND FERRY COMPANY

Address ***Company*** Shell Bay, Studland, Swanage, Dorset BH19 3BA. **Tel** +44 (0)1929 450203, **Fax** +44 (0)1929 450498), ***Ferry*** Floating Bridge, Ferry Way, Sandbanks, Poole, Dorset BH13 7QN. **Tel** +44 (0)1929 450203.

Internet Email email@sandbanksferry.co.uk **Website** www.sandbanksferry.co.uk *(English)*

Route Sandbanks – Studland (Dorset).

1	BRAMBLE BUSH BAY	93	74.4m	400P	48C	BA

BRAMBLE BUSH BAY chain ferry, built by Richard Dunston (Hessle) Ltd, Hessle, UK for the *Bournemouth-Swanage Motor Road and Ferry Company.*

CUMBRIA COUNTY COUNCIL

Address Community, Economy & Environment Department, The Courts, Carlisle CA3 8NA. **Tel** +44 (0)1228 606744, **Fax** +44 (0)1228 514974.

Internet Email john.robinson@cumbriacc.gov.uk

Route Bowness-on-Windermere – Far Sawrey.

1	MALLARD	90	25.9m	140P	18C	BA

MALLARD Chain Ferry built by F L Steelcraft, Borth, Dyfed for *Cumbria County Council.*

DARTMOUTH – KINGSWEAR FLOATING BRIDGE CO LTD

Address Dart Marina, Sandquay Road, Dartmouth, Devon TQ6 9PH. **Tel** +44 (0)1803 833351.

Route Dartmouth – Kingswear (Devon) across River Dart (higher route) (forms part of A379).

1	HIGHER FERRY	60	42.7m	136P	18C	BA

HIGHER FERRY Built by Philip Ltd, Dartmouth, UK. Diesel electric paddle propelled vessel guided by cross-river cables.

ISLE OF WIGHT COUNCIL (COWES FLOATING BRIDGE)

Address Ferry Office, Medina Road, Cowes, Isle of Wight PO31 7BX. **Tel** +44 (0)1983 293041.

Route Cowes – East Cowes.

1	NO 5	76	33.5m	-	15C	BA

NO 5 Chain ferry built by Fairey Marine, East Cowes, UK for *Isle of Wight County Council*, now *Isle of Wight Council.*

KING HARRY STEAM FERRY COMPANY

Address Feock, Truro, Cornwall TR3 6QJ. **Tel** +44 (0)1872 862312, **Fax** +44 (0)1872 863355.

Internet Email info@kingharry.fq.co.uk **Website** www.kingharry-info.co.uk *(English)*

Route Across River Fal, King Harry Ferry (Cornwall).

1	KING HARRY FERRY	74	44.2m	100P	23C	BA

KING HARRY FERRY Chain ferry built by Dredge-Marine Ltd, Falmouth, UK for *King Harry Steam Ferry Company.*

On order

2	NEWBUILDING	06	55.2m	150P	34C	BA

NEWBUILDING On order from Pendennis Shipyard, Falmouth to replace the existing ferry.

REEDHAM FERRY

Address Reedham Ferry, Ferry Inn, Reedham, Norwich NR13 3HA. **Tel** +44 (0)1493 700429, **Fax** +44 (0)1493 700999.

Route Acle – Reedham – Norton (across River Yare, Norfolk).

1	REEDHAM FERRY	84	11.3m	12P	3C	BA

REEDHAM FERRY Chain ferry built by Newsons, Oulton Broad, UK for *Reedham Ferry.* Maximum weight, 12 tons.

SOUTH HAMS DISTRICT COUNCIL

Address Lower Ferry Office, The Square, Kingswear, Dartmouth, Devon TQ6 0AA. **Tel** +44 (0)1803 752342, **Fax** +44 (0)1803 752227.

Route Dartmouth – Kingswear (Devon) across River Dart (lower route).

1	THE TOM AVIS	94	33.5m	50P	8C	BA
2	THE TOM CASEY	89	33.5m	50P	8C	BA

THE TOM AVIS Float propelled by tugs built by Alan Toms, Fowey, UK for *South Hams District Council.*

THE TOM CASEY Float propelled by tugs built Cozens, Portland, UK for *South Hams District Council.*

TORPOINT FERRY

Address 2 Ferry Street, Torpoint, Cornwall PL11 2AX. Tel +44 (0)1752 812233, Fax +44 (0)1752 816873.

INTERNET Website www.torpointferry.org.uk *(English)*

Route Devonport (Plymouth) – Torpoint (Cornwall) across the Tamar. The three ferries operate in parallel, each on its own 'track'. Pre-booking is not possible and the above number cannot be used for that purpose.

1●	LYNHER	-	61	-	70.7m	350P	48C	-	BA	-
2	LYNHER II	850t	05	-	73.0m	350P	73C	-	BA	UK
3●	PLYM	-	68	-	70.7m	350P	54C		BA	
4	PLYM II	850t	04	-	73.0m	350P	73C	-	BA	UK
5●	TAMAR	-	60	-	70.7m	350P	48C	-	BA	
6	TAMAR II	850t	05	-	73.0m	350P	73C	-	BA	UK

LYNHER, TAMAR Chain ferries built by Thorneycroft, Southampton, UK. In 2005 withdrawn and laid up.

LYNHER II, PLYM II, TAMAR II Chain ferries built by Ferguson Shipbuilders Ltd, Port Glasgow, UK to replace 1960s built ships. Unlike other chain ferries they are registered as 'Passenger/Ro-Ro Cargo' ships and thus have gross tonnage and nation of registry.

PLYM Chain ferry built by Charles Hill & Son Ltd, Bristol. In 2004 withdrawn and laid up.

WATERFORD CASTLE HOTEL

Address The Island, Waterford, Irish Republic. Tel +353 (0)51 78203.

Internet Email info@waterfordcastle.com **Website** www.waterfordcastle.com *(English (mainly about hotel; little about ferry))*

Route Grantstown – Little Island (in River Suir, County Waterford).

1	LITTLE ISLAND FERRY	68	-	24P	6C	BA

LITTLE ISLAND FERRY Chain ferry built at Cork, Irish Republic for *Waterford Castle Hotel*.

There are a surprisingly large number of passenger only ferries operating in the British Isles, mainly operated by launches and small motor boats. There are, however, a few 'major' operators who operate only passenger vessels (of rather larger dimensions) and have not therefore been mentioned previously.

section 5 major passenger only ferries

gb & ireland

6° West (trading name of West Highland Seaways) SPIRIT OF SKYE (2004, 36 passengers). **Route operated** Gairloch – Portree (Skye). **Tel** +44 (0)0800 328 6426, **Email** info@overtheseatoskye.com **Website** www.overtheseatoskye.com *(English)*

Clyde Marine Services CRUISER (119t, 1974, 24.4m, 249 passengers (ex POOLE SCENE, 2001), FENCER (18t, 1976, 11.0m, 33 passengers), KENILWORTH (44t, 1936, 18.3m, 126 passengers (ex HOTSPUR II (Southampton – Hythe ferry) 1979)), ROVER (48t, 1964, 19.8m, 84 passengers), THE SECOND SNARK (45t, 1938, 22.9m, 120 passengers). **Routes operated** Gourock – Kilcreggan – Helensburgh (generally the KENILWORTH), Gourock – Kilcreggan – Helensburgh – Dunoon – Rothesay – Millport – Lochranza and other Clyde coast ports (generally THE SECOND SNARK). **Tel** +44 (0)1475 721281, **Fax** +44 (0)1475 888023, **Websites** www.clyde-marine.co.uk www.secondsnark.co.uk *(English)*

Dart Pleasure Craft DARTMOUTH PRINCESS (22t, 1990, 18.3m, 156 passengers, KINGSWEAR BELLE (43t, 1972, 18.0m, 257 passengers); **Route operated** Dartmouth – Kingswear. EDGCUMBE BELLE (35t, 1957, 17.7m, 150 passengers), **Route operated** Dartmouth – Greenway (summer only). **Note** Pleasure craft owned by this operator are also used for the ferry service on some occasions. **Tel** +44 (0)1803 834488, **Fax** +44 (0)1803 835248, **Email** sales@riverlink.co.uk **Website** www.riverlink.co.uk *(English)*

Doolin Ferry Company/O'Brien Shipping DONEMARK (70t, 1978, 19.8m, 65 pass), HAPPY HOOKER (77t, 1989, 19.8m, 96 passengers), QUEEN OF ARAN (113t, 1976, 20.1m, 96 passengers), ROSE OF ARAN (113t, 1976, 20.1m, 96 passengers), TRANQUILITY (62t, 1988, 15.8m, 96 passengers). **Route operated** Doolin – Inishere, Doolin – Inishmaan, Doolin – Inishmore. OILEAN ARANN (416t, 1992, 39.6m, 190 passengers). **Route operated** Galway – Inishere, Galway –

Inishmaan, Galway – Inishmore. **Tel** +353 (0)65 7074455, **Fax** +353 (0)65 7074417, **Email** doolinferries@eircom.net **Web Site** www.doolinferries.com *(English)*

G&T Ferries (trading name of Lower Thames & Medway Passenger Boat Co Ltd) DUCHESS M (71t, 1956, 23.8m, 124 passengers) (ex VESTA 1979), PRINCESS POCAHONTAS (180t, 1962, 33.5m, 207 passengers (ex FREYA II 1989, LABOE I 1985, LABOE 1984 (excursion vessel operating regularly to Greenwich, Westminster and Chelsea also occasionally to Rochester, Southend and Whitstable but sometimes covers the ferry roster)). **Route operated** Gravesend (Kent) – Tilbury (Essex), **Tel** +44 (0)1732 353448, **Direct Line to Ferry** +44 (0)7973 390124, **Email** enquiry@princess-pocahontas.com **Web Site** www.princess-pocahontas.com *(English)*

Gosport Ferry GOSPORT QUEEN (159t, 1966, 30.5m, 250 passengers), PORTSMOUTH QUEEN (159t, 1966, 30.5m, 250 passengers), SOLENT PRINCE (12t, 1981, 43m, 60 passengers) (ex JENNY ANN, ex WATER WYTCH) (mainly used on charter work), SPIRIT OF GOSPORT (250t, 2001, 32.6m, 247 passengers). **Under construction** SPIRIT OF PORTSMOUTH (2005, 32.6m, 300 passengers). **Route operated** Gosport – Portsmouth. **Tel** +44 (0)23 9252 4551, **Email** admin@gosportferry.co.uk **Web Site** www.gosportferry.co.uk *(English)*

Hovertravel FREEDOM 90 (1990, 25.4m, 98 passengers) (BHC AP1-88/100S hovercraft (converted from AP1-88/100 in 2000)), ISLAND EXPRESS (1985, 25.4m, 98 passengers) (BHC AP1-88/100S hovercraft (converted from BHC AP1-88/100 in 2001)) (ex FREJA VIKING 2002). **Route operated** Southsea – Ryde. **Tel** +44 (0)1983 811000, **Fax** +44 (0)1983 812859, **Email** info@hovertravel.co.uk, **Website** www.hovertravel.co.uk *(English)*

Inis Mór Ferries QUEEN OF ARAN II (2001, 29.9m, 227 passengers), **Route operated** Rossaveal (Co Galway) – Inis Mór (Aran Islands). **Tel** +353 (0)91 566 535, **Fax** +353 (0)91 534315, **E-mail** info@queenofaran2.com, **Website** www.queenofaran2.com *(English, French, German, Italian, Spanish)*

Island Ferries ARAN EXPRESS (117t, 1984, 27.4m, 180 passengers), ARAN FLYER (170t, 1988, 33.5m, 208 passengers), ARAN SEABIRD (164t, 1976, 27.7m, 181 passengers), CEOL NA FARRAIGE (200t, 2001, 35.4m, 294 passengers), DRAÍOCHT NA FARRAIGE (200t, 1999, 35.4m, 294 passengers). **Routes operated** Rossaveal (Co Galway) – Aran Islands. **Tel** +353 (0)91 568903 (572273 after 19.00), **Fax** +353 (0)91 568538, **Email** island@iol.ie, **Website** www.aranislandferries.com *(English)*

John O'Groats Ferries PENTLAND VENTURE (186t, 1987, 29.6m, 250 passengers). **Route operated** John O'Groats – Burwick (Orkney). **Tel** +44 (0)1955 611353, **Email** Office@jogferry.co.uk **Website** www.jogferry.co.uk *(English)*

Lundy Company OLDENBURG (288t, 1958, 43.6m, 267 passengers). **Routes operated** Bideford – Lundy Island, Ilfracombe – Lundy Island. Also North Devon Coastal Cruises. **Tel** +44 (0)1237 470074, **Fax** +44 (0)1237 477779, **Email** info@lundyisland.co.uk **Web Site** www.lundyisland.co.uk *(English)*

Manche Iles Express (trading name of Société Morbihannaise de Navigation) VICTOR HUGO (ex SALTEN 2003) (387t, 1997, 35.0m, 190 passengers) (catamaran), **Routes operated** Portbail or Carteret – Jersey, Guernsey and Sark, Dielette – Alderney – Guernsey, MARIN MARIE (ex AREMETI 3 2003) (608t, 1994, 40.0m, 356 passengers). **Route operated** Granville – Jersey. **Tel** ***Jersey*** +44 (0)1534 880 756, ***Guernsey*** +44 (0)1481 701 316, ***Granville*** +33 2 33 610 888, ***Carteret*** +33 2 33 011 011, ***Web Site:*** www.manche-iles-express.com.

Mersey Ferries ROYAL DAFFODIL (ex OVERCHURCH 1999) (468t, 1962, 46.6m, 860 passengers), ROYAL IRIS OF THE MERSEY (ex MOUNTWOOD 2002) (464t, 1960, 46.3m, 750 passengers), SNOWDROP (ex WOODCHURCH 2004) (464t, 1960, 46.6m, 750 passengers). **Routes operated** Liverpool (Pier Head) – Birkenhead (Woodside), Liverpool – Wallasey (Seacombe). Also regular summer cruises from Pier Head to Salford along the Manchester Ship Canal. **Tel** *Admin* +44 (0)151 639 0609, ***Reservations*** +44 (0)151 330 1444, **Fax** +44 (0)151 639 0578, **Email** info@merseyferries.co.uk **Website** www.merseyferries.co.uk *(English)*

Nexus (trading name of Tyne & Wear PTE) PRIDE OF THE TYNE (222t, 1993, 24.0m, 240 passengers), SHIELDSMAN (93.2t, 1976, 24.0m, 350 passengers). **Route operated** North Shields

– South Shields. Also cruises South Shields – Newcastle. **Tel** +44 (0)191 454 8183, **Fax** +44 (0)191 427 9510, **Web Site** www.nexus.org.uk *(English)*

Strathclyde Passenger Transport (trading name of Strathclyde PTE) RENFREW ROSE (65t, 1984, 21.9m, 50 passengers), YOKER SWAN (65t, 1984, 21.9m, 50 passengers). **Route operated** Renfrew – Yoker. **Note** although this a passenger only service, the vessels are built as small front loading car ferries and are able to convey one vehicle if necessary. This facility is sometimes used for the conveyance of ambulances. **Tel** +44 (0)141 885 2123, **Fax** +44 (0)141 432 1025, **Email** liz.parkes@spt.co.uk **Website** www.spt.co.uk/Travel/ferries.html *(English)*

Thames Clippers (part of Collins River Enterprises) ABEL MAGWITCH (25.6t, 1999, 18.3m, 60 passengers (tri-maran)), HURRICANE CLIPPER (181t, 2002, 37.8m, 27.5k, 220 passengers), SKY CLIPPER (60t, 1992, 25.0m, 62 passengers) (ex VERITATUM 1995, SD10 2000), STORM CLIPPER (60t, 1992, 25.0m, 62 passengers) (ex DHL WORLDWIDE EXPRESS 1995, SD11 2000), TWIN STAR (27t,1974,19.2m,200 passengers). **Routes operated** Savoy Pier (Embankment) – Canary Wharf, Canary Wharf – Rotherhithe (Hilton Hotel) (usually the TWIN STAR). **Tel** +44 (0)20 7977 6892, **Fax** +44(0) 20 7481 8300, **Email** sean@thamesclippers.com **Website** www.thamesclippers.com *(English)*

Waverley Excursions BALMORAL (735t, 1949, 62.2m, 800 passengers), WAVERLEY (693t, 1947, 73.2m, 950 passengers). **Routes operated** Excursions all round British Isles. However, regular cruises in the Clyde, Bristol Channel, South Coast and Thames provide a service which can be used for transport purposes and therefore both vessels are, in a sense, ferries. The WAVERLEY is the only seagoing paddle steamer in the world. **Tel** +44 (0)845 130 4647, **Fax** +44 (0)141 248 2150, **Email** info@waverleyexcursions.co.uk **Website** www.waverleyexcursions.co.uk *(English)*

White Horse Ferries GREAT EXPECTATIONS (66t, 1992, 21.3m, 162 passengers) (catamaran), HOTSPUR IV (50t, 1946, 19.5m, 125 passengers). **Route operated** Southampton – Hythe (Hants). *Head Office* **Tel**. +44 (0)1793 618566, **Fax** +44 (0)1793 488428, *Local Office* **Tel** +44 (0)23 8084 0722, **Fax** +44 (0)23 8084 6611, **Email** post@hytheferry.co.uk **Website** www.hytheferry.co.uk *(English)*

Twin Star *(Nick Widdows)*

EXPRESS
Stena Line
EXPRESS

Stena Carisma & Stena Germanica *(Miles Cowsill)*

section 6 major passenger ferries

northern europe

BASTØ FOSEN

THE COMPANY *Bastø Fosen* is a Norwegian private sector company, a subsidiary of *Fosen Trafikklag* of Trondheim.

MANAGEMENT Managing Director Olav Brein, **Operations Manager** Kirsti Been Tofte.

ADDRESS PO Box 94, 3191 Horten, Norway.

TELEPHONE Administration +47 33 03 17 40, **Reservations** not applicable.

FAX Administration & Reservations +47 33 03 17 49.

INTERNET Email basto@fosen.no **Website** www.basto-fosen.no *(Norwegian)*

ROUTE OPERATED Moss Horten (across Oslofjord, Norway) (30 mins; *(1,2,3)*; up to every 30 mins).

1	BASTØ I	5505t	97	14.0k	109.0m	500P	200C	18L	BA	NO
2	BASTØ II	5505t	97	14.0k	109.0m	500P	200C	18L	BA	NO
3	BASTØ III	5500t	05	18.0k	116.2m	550P	212C	18L	BA	NO

BASTØ I, BASTØ II Built by Fosen Mekaniske Verksted, Frengen, Norway for *Bastø Fosen.*

BASTØ III Built by Stocznia Remontowa, Gdansk, Poland for *Bastø Fosen.*

BORNHOLMSTRAFIKKEN

THE COMPANY *BornholmsTrafikken* is a Danish public sector joint-stock company.

MANAGEMENT Managing Director Mads Kofod, **Sales and Marketing Manager** Ole B Larsen.

ADDRESS Havnen, 3700 Rønne, Denmark.

TELEPHONE Administration +45 56 95 18 66, **Reservations** +45 56 95 18 66.

FAX Administration & Reservations +45 56 91 07 66.

INTERNET Website www.bornholmstrafikken.dk *(Danish, Swedish, English)*

ROUTES OPERATED Conventional Ferries Rønne (Bornholm, Denmark) – Køge (6 hrs 30 mins; *(1,2)*; 1 per day), Rønne – Ystad (Sweden) (2 hrs 40 mins; *(1,2)*; 1 or 2 per day), ***Summer only:*** Rønne – Sassnitz (Germany) (3 hrs 30 mins; (1,2,*3*); 1 or 2 per day), Køge - Sassnitz (5hrs 30mins;(1,2); 1 per day). **Fast Ferry** Rønne – Ystad (Sweden) (1 hr 20 mins; *(4)*; up to 5 per day).

1	DUEODDE	12504t	05	18.5k	124.9m	400P	342C	88T	A	DK
2	HAMMERODDE	12504t	05	18.5k	124.9m	400P	342C	88T	A	DK
3	POVL ANKER	12131t	78	19.5k	121.0m	1500P	262C	26T	BA	DK
4»	VILLUM CLAUSEN	6402t	99	40.0k	86.6m	1065P	215C	-	BA	DK

DUEODDE Built by Volharding Shipyards, Westerbroek, Netherlands. Due to enter service May 2005.

HAMMERODDE Built by Merwede Shipyard, Hardinxveld-Giessendam, Netherlands. Due to enter service May 2005.

Povl Anker (*FotoFlite*)

Gotland (*Miles Cowsill*)

POVL ANKER Built by Aalborg, Denmark for *BornholmsTrafikken*. Used on the Rønne – Copenhagen, Rønne – Ystad and (until December 2002) Rønne – Sassnitz services. In 2005 to restore the Rønne – Sassnitz and act as reserve vessel.

VILLUM CLAUSEN Austal Auto-Express 86 catamaran built at Fremantle, Australia for *BornholmsTrafikken*. Used on the Rønne – Ystad service. Car capacity increased in 2005.

COLOR LINE

THE COMPANY *Color Line ASA* is a Norwegian private sector stock-listed limited company. The company merged with *Larvik Scandi Line* of Norway (which owned *Larvik Line* and *Scandi Line*) in 1996. *Larvik Line's* operations were incorporated into *Color Line* in 1997; *Scandi Line* continued as a separate subsidiary until 1999, when it was also incorporated into *Color Line*. The marketing name *Color Scandi Line* was dropped at the end of 2000.

MANAGEMENT Managing Director Trond Kleivdal, **Marketing Manager** Elisabeth Anspach.

ADDRESS *Commercial* Postboks 1422 Vika, 0115 OSLO, Norway, ***Technical Management*** Color Line Marine AS, PO Box 2090, 3210 Sandefjord, Norway.

TELEPHONE Administration +47 22 94 44 00, **Reservations** +47 810 00 811.

FAX Administration +47 22 83 04 30, **Reservations** +47 22 83 07 76.

INTERNET Website www.colorline.com *(Norwegian, Danish, English, Swedish, German)*

ROUTES OPERATED Conventional Ferries *All year* Oslo (Norway) – Kiel (Germany) (19 hrs 30 mins; *(3,7)*; 1 per day), Oslo – Hirtshals (Denmark) (8 hrs 30 mins; *(4)*; 1 per day), Kristiansand (Norway) – Hirtshals (4 hrs 30 mins; *(2)*; 2 per day), Larvik (Norway) – Frederikshavn (Denmark) (6 hrs 15 mins; *(5 (summer peak only), 8)*; up to 3 per day), Sandefjord (Norway) – Strömstad (Sweden) (2 hrs 30 mins; *(1, 6)*; up to 6 per day), Bergen (Norway) – Stavanger (Norway) – Hirtshals (18 hrs; *(9)*; 3 per week (extra sailing Hirtshals – Stavanger only in summer peak)). ***Not peak summer period*** Larvik (Norway) – Hirtshals (5 hrs 30 mins; *(5)*; 1 per day). **Fast Ferry (under the name 'Color Line Express') Summer only** Kristiansand – Hirtshals (2 hrs 30 mins; *(10)*; 3 per day).

1	BOHUS	9149t	71	20.0k	123.4m	1165P	240C	34T	BA	NO
2	CHRISTIAN IV	21699t	81	19.0k	153.4m	1860P	530C	56T	BAS2	NO
3	COLOR FANTASY	74027t	04	22.3k	224.0m	2750P	750C	90T	BA2	NO
4	COLOR FESTIVAL	34417t	85	22.0k	168.0m	2000P	650C	80T	BA	NO
5	COLOR TRAVELLER	17046t	81	17.0k	140.8m	1100P	750C	84T	BAS2	NO
6	COLOR VIKING	19763t	85	17.0k	134.0m	2000P	320C	40T	BA2	NO
7	KRONPRINS HARALD	31914t	87	23.5k	166.3m	1458P	730C	90T	BA	NO
8	PETER WESSEL	30316t	81	19.0k	168.5m	2138P	562C	136T	BA	NO
9	PRINSESSE RAGNHILD	35855t	81	22.0k	205.3m	1530P	600C	70T	BA	NO
10»	SILVIA ANA L	7895t	96	41.0k	125.0m	1065P	200C	4L	A	BS

BOHUS Built by Aalborg Værft A/S, Aalborg, Denmark as the PRINSESSAN DESIREE for *Rederi AB Göteborg-Frederikshavn Linjen* of Sweden (trading as *Sessan Linjen*) for their service between Gothenburg and Frederikshavn. In 1981 the company was taken over by *Stena Line* and she became surplus to requirements. During 1981 she had a number of charters including *B&I Line* of Ireland and *Sealink* UK. In 1982 she was chartered to *Sally Line* to operate as second vessel on the Ramsgate – Dunkerque service between June and September. She bore the name VIKING 2 in large letters on her hull although she was never officially renamed and continued to bear the name PRINSESSAN DESIREE on her bow and stern. In September 1982 she returned to *Stena Line* and in 1983 she was transferred to subsidiary company *Varberg-Grenaa Line* for their service between Varberg (Sweden) and Grenaa (Denmark) and renamed the EUROPAFÄRJAN. In 1985 she was renamed the EUROPAFÄRJAN II. In 1986, following a reorganisation within the *Stena Line* Group, ownership was transferred to subsidiary company *Lion Ferry AB* and she was named the LION PRINCESS. In 1993 she was sold to *Scandi Line* and renamed the BOHUS. In 1999 *Scandi Line* operations were integrated into *Color Line*.

CHRISTIAN IV Built by AG Weser Seebeckswerft, Bremerhaven, Germany as the OLAU BRITANNIA for *Olau Line* of Germany for their service between Vlissingen (Netherlands) and Sheerness (England). In 1989 sold to *Nordström & Thulin* of Sweden for delivery in spring 1990. She was subsequently resold to *Fred. Olsen Lines* of Norway and, on delivery, renamed the BAYARD and used on their service between Kristiansand and Hirtshals. In December 1990 she was acquired by *Color Line* and in 1991 renamed the CHRISTIAN IV. She continues to operate on that route.

COLOR FANTASY Built by Kvaerner Masa-Yard, Turku Finland for *Color Line* to replace the PRINSESSE RAGNHILD on the Oslo – Kiel service.

COLOR FESTIVAL Built by Oy Wärtsilä AB, Helsinki, Finland as the SVEA for *Johnson Line* for the *Silja Line* Stockholm – Mariehamn – Turku service. During winter 1991/92 she was extensively rebuilt and in 1991 renamed the SILJA KARNEVAL; ownership was transferred to *Silja Line*. In 1993 she was sold to *Color Line* and renamed the COLOR FESTIVAL. She is used on the Oslo – Hirtshals service.

COLOR TRAVELLER Built by Oy Wärtsilä AB, Helsinki, Finland as the TRAVEMÜNDE for *Gedser-Travemünde Ruten* of Denmark for their service between Gedser (Denmark) and Travemünde (Germany). In 1986 the company's trading name was changed to *GT Linien* and in 1987, following the take-over by *Sea-Link AB* of Sweden, it was further changed to *GT Link*. The vessel's name was changed to the TRAVEMÜNDE LINK. In 1988 she was purchased by *Rederi AB Gotland* of Sweden, although remaining in service with *GT Link*. Later in 1988 she was chartered to *Sally Ferries* and entered service in December on the Ramsgate – Dunkerque service. She was renamed the SALLY STAR. In 1997 she was transferred to *Silja Line*, to operate between Vaasa and Umeå during the summer period and operated under the marketing name WASA EXPRESS (although not renamed). She returned to *Rederi AB Gotland* in autumn 1997, was renamed the THJELVAR and entered service with *Destination Gotland* in January 1998. Withdrawn and laid up in December 2003. In 2004 chartered to *Color Line* to inaugurate a new service between Larvik and Hirtshals. Renamed the COLOR TRAVELLER. Operates in reduced passenger mode on this service but in summer peak period operates between Frederikshavn and Larvik in full passenger mode.

COLOR VIKING Built by Nakskov Skipsværft, Nakskov, Denmark as the PEDER PAARS for *DSB (Danish State Railways)* for their service between Kalundborg (Sealand) and Århus (Jutland). In 1990 purchased by *Stena Line* of Sweden for delivery in 1991. In 1991 renamed the STENA INVICTA and entered service on the *Sealink Stena Line* Dover – Calais service. She was withdrawn from the route in February 1998, before the formation of *P&O Stena Line* but ownership was transferred to that company. In summer 1998, she was chartered to *Silja Line* to operate between Vaasa and Umeå under the marketing name 'WASA JUBILEE'. In autumn 1998 she was laid up at Zeebrugge. She remained there until autumn 1999 when she was chartered to *Stena Line* to operate between Holyhead and Dublin. In 2000 she was chartered to *Color Line* and renamed the COLOR VIKING and in April entered service on the Sandefjord – Strömstad service. In 2002 purchased by *Color Line*.

KRONPRINS HARALD Built by Oy Wärtsilä AB, Turku, Finland for *Jahre Line* of Norway for the Oslo – Kiel service. In 1991 ownership was transferred to *Color Line*.

PETER WESSEL Built by Öresundsvarvet AB, Landskrona, Sweden for *Rederi AB Gotland* of Sweden. A sister vessel of the VISBY (see *Destination Gotland*), it was intended that she should be named the GOTLAND. However, she was delivered as the WASA STAR and chartered to *Vaasanlaivat* of Finland and used on their Vaasa – Sundsvall service. In 1982 she was chartered to *Karageorgis Line* of Greece for service between Patras (Greece) and Ancona (Italy). This charter was abruptly terminated in 1983 following a dispute over payment of charter dues. She returned the Baltic and was laid up until February 1984 when she was sold to *Larvik Line*. She was renamed the PETER WESSEL. In 1988 she was lengthened. In 1996 acquired by *Color Line*. She remains on the Larvik – Moss – Frederikshavn route.

PRINSESSE RAGNHILD Built by Kiel Howaldtswerke, Kiel, Germany for *Jahre Line* of Norway for the Oslo – Kiel service. In 1991 ownership transferred to *Color Line*. In 1992 rebuilt in Spain with an additional mid-ships section and additional decks. In 2005 started new Bergen – Stavanger – Hirtshals service.

SILVIA ANA L Bazan Alhambra monohull vessel built at San Fernando, Spain for *Buquebus* of

Argentina. Initially operated between Buenos Aires (Argentina) and Piriapolis (Uruguay). In 1997 chartered to *Color Line* to operate between Kristiansand and Hirtshals. During winter 1997/98 she again operated in South America but returned to *Color Line* in spring 1998. This was repeated during winters 1998/9 and 1999/2000 but during winter 2000/2001 she remained laid up in Europe. In 2001 she was sold to *MDFC Aircraft* of the Irish Republic and chartered to *Color Line* for four years. In 2005 purchased by *Color Line*.

Under Construction

11	NEWBUILDING	-	07	22.3k	-	-	-	-	BA2	NO

NEWBUILDING To be built by Aker Yards, Rauma, Finland for the Oslo – Kiel route. Letter of intent signed in February 2005; contract expected to be signed in June 2005.

DESTINATION GOTLAND

THE COMPANY *Destination Gotland AB* is a Swedish private sector company owned by *Rederi AB Gotland*. It took over the operations of services to Gotland from 1st January 1998 on a six-year concession. Originally jointly owned by *Rederi AB Gotland* and *Silja Line*, *Silja Line* involvement in the company ceased at the end of 1998.

MANAGEMENT Managing Director Sten-Crister Forsberg, **Marketing Manager** Per-Erling Evensen.

ADDRESS PO Box 1234, 621 23 Visby, Gotland, Sweden.

TELEPHONE Administration +46 (0)498-20 18 00, **Reservations** +46 (0)771-22 33 00.

FAX Administration & Reservations +46 (0)498-20 13 90.

INTERNET Email info@destinationgotland.se **Website** www.destinationgotland.se *(Swedish, English)*

ROUTES OPERATED Fast Conventional Ferries Visby (Gotland) – Nynäshamn (Swedish mainland) (3 hrs 10 mins; *(1,3)*; 1/2 per day), Visby – Oskarshamn (Swedish mainland) (2 hrs 50 mins; *(1,3)*; 1/2 per day). **Fast Ferry (summer only)** Visby – Nynäshamn (2 hrs 50 mins; *(2)*; up to 2 per day), Visby – Oskarshamn (Swedish mainland) (2 hrs 30 mins; *(2)*; occasional).

1	GOTLAND	29000t	03	28.5k	195.8m	1500P	500C	118T	BA	SW
2»	GOTLANDIA	5632t	99	35.0k	112.5m	700P	140C	-	A	SW
3	VISBY	29000t	03	28.5k	195.8m	1500P	500C	118T	BA	SW

GOTLAND Built by Guangzhou Shipyard International, Guangzhou, China for *Rederi AB Gotland* for use on *Destination Gotland* services.

GOTLANDIA Alstom Leroux Corsair 11500 monohull vessel built at Nantes, France as the GOTLAND for *Rederi AB Gotland* and chartered to *Destination Gotland*. In 2003 renamed the GOTLANDIA.

VISBY Built by Guangzhou Shipyard International, Guangzhou, China for *Rederi AB Gotland* and used on *Destination Gotland* services.

Under Construction

4»	NEWBUILDING	-	06	40.0k	122.0m	900P	160C	-	A	SW

NEWBUILDING Fincantieri SF700 monohull fast ferry under construction at Riva Trigoso, Italy for *Rederi AB Gotland* for use by *Destination Gotland*.

DFDS SEAWAYS

THE COMPANY *DFDS Seaways* is the passenger division of *DFDS A/S*, a Danish private sector company.

MANAGEMENT Managing Director DFDS A/S Ole Frie, **Director DFDS Seaways** Søren Jespersen.

ADDRESS Sundkrogsgåde 11, 2100 Copenhagen Ø, Denmark.

TELEPHONE Administration +45 33 42 33 42, **Reservations** +45 33 42 30 00.

FAX Administration & Reservations +45 33 42 33 41.

INTERNET Website www.dfdsseaways.com *(Danish, Dutch, English, German, Norwegian, Swedish)*

ROUTE OPERATED Copenhagen – Helsingborg (Sweden) – Oslo (Norway) (16 hrs; *(1,2)*; 1 per day). See Section 1 for services operating to Britain.

1	CROWN OF SCANDINAVIA	35498t	94	22.0k	169.4m	1940P	450C	50T	BA	DK
2	PEARL OF SCANDINAVIA	40039t	89	21.0k	178.4m	2090P	350C	70T	BA	DK

CROWN OF SCANDINAVIA Launched by Brodogradevna Industrija, Split, Croatia for *Euroway* for their Lübeck – Travemünde – Malmö service as the THOMAS MANN. However, political problems led to serious delays and, before delivery, the service had ceased. She was purchased by *DFDS*, renamed the CROWN OF SCANDINAVIA and introduced onto the Copenhagen – Oslo service.

PEARL OF SCANDINAVIA Built by Wärtsilä Marine, Turku, Finland as the ATHENA for *Rederi AB Slite* of Sweden (part of *Viking Line*) and used on 24 hour cruises from Stockholm to Mariehamn (Åland). In 1993 the company went into liquidation and she was sold to *Star Cruises* of Malaysia for cruises in the Far East. She was renamed the STAR AQUARIUS. Later that year she was renamed the LANGKAPURI STAR AQUARIUS. In February 2001 sold to *DFDS* and renamed the AQUARIUS. After rebuilding, she was renamed the PEARL OF SCANDINAVIA and introduced onto the Copenhagen – Oslo service.

REDERIJ DOEKSEN

THE COMPANY *Doeksen Transport Group bv* is a Dutch private sector company. Ferries are operated by subsidiary *Terschellinger Stoomboot Maatschappij*, trading as *Rederij Doeksen*.

MANAGEMENT Managing Director P Melles, **Marketing Manager** C Dekker.

ADDRESS Willem Barentskade 21, Postbus 40, 8880 AA West Terschelling, Netherlands.

TELEPHONE Administration +31 (0)562 442141, **Reservations** +31 (0)562 446111.

FAX Administration & Reservations +31 (0)562 443241.

INTERNET Email info@rederij-doeksen.nl **Website** www.rederij-doeksen.nl *(Dutch)*

ROUTES OPERATED Conventional Ferries Harlingen (Netherlands) – Terschelling (Frisian Islands) (2 hrs; *(1,3)* *(*up to 6 per day), Harlingen – Vlieland (Frisian Islands) (1 hr 45 mins; *(5,7)*; 3 per day), **Fast Passenger Ferries** Harlingen – Terschelling (45 mins; *(2,4)*; 2 to 3 per day), Harlingen – Vlieland (45 mins; *(2,4)*; 2 per day), Vlieland – Terschelling (30 mins; *(2,4)*; 2 per day).

1	FRIESLAND	3583t	89	14.0k	69.0m	1750P	122C	12L	BA	NL
2»p	KOEGELWIECK	439t	92	33.0k	36.7m	317P	0C	0L	-	NL
3	MIDSLAND	1812t	74	15.5k	77.9m	1200P	55C	6L	BA	NL
4»p	NAJADE	164t	99	32.0k	31.8m	184P	0C	0L	-	NL
5	OOST-VLIELAND	1350t	70	15.0k	62.6m	1100P	45C	4L	BA	NL
6F	NOORD-NEDERLAND	-	02	14.0k	48.0m	12P	-	9L	BA	NL
7	VLIELAND	-	05	15.0k	64.0m	1300P	58C	-	BA	NL

FRIESLAND Built by Van der Giessen-de Noord, Krimpen aan den IJssel, Rotterdam, Netherlands for *Rederij Doeksen*. Used on the Harlingen – Terschelling route.

KOEGELWIECK Harding 35m catamaran built at Rosendal, Norway for *Rederij Doeksen* to operate between Harlingen and Terschelling, Harlingen and Vlieland and Terschelling and Vlieland.

MIDSLAND Built by Werftunion GmbH & Co, Cassens-Werft, Emden, Germany as the RHEINLAND for *AG Ems* of Germany. In 1993 purchased by *Rederij Doeksen* and renamed the MIDSLAND. Used mainly on the Harlingen – Terschelling route but also used on the Harlingen – Vlieland service. She is now a reserve vessel.

NAJADE SBF Shipbuilders 31m monohull built at Henderson, Australia for *Rederij Doeksen* to operate between Harlingen and Terschelling, Harlingen and Vlieland and Terschelling and Vlieland.

NOORD-NEDERLAND Catamaran built by ASB, Harwood, New South Wales, Australia for *Rederij Doeksen*. Used on freight services from Harlingen to Terschelling and Vlieland.

OOST-VLIELAND Built by C Cassens Schiffswerft, Emden, Germany as the OSTFRIESLAND for *AG Ems* of Germany. In 1981 purchased by *Rederij Doeksen* and renamed the SCHELLINGERLAND. In 1994 renamed the OOST VLIELAND. Now mainly used on the Harlingen – Vlieland service.

VLIELAND Catamaran built by FBMA Babcock Marine, Cebu, Philippines for *Rederij Doeksen* to operate between Harlingen and Vlieland.

ECKERÖ LINE

THE COMPANY *Eckerö Line Ab Oy* is a Finnish company, 100% owned by *Eckerö Linjen* of Åland, Finland. Until January 1998, the company was called *Eestin-Linjat*.

MANAGEMENT Managing Director Jarl Danielsson, **Marketing Director** Håkan Nordström.

ADDRESS Hietalahdenranta 13, 00180 Helsinki, Finland.

TELEPHONE Administration +358 (0)9 22885421, **Reservations** +358 (0)9 2288544.

FAX Administration & Reservations +358 (0)9 22885222.

INTERNET Email info@eckeroline.fi **Website** www.eckeroline.fi *(Swedish, Finnish, English, German)*

ROUTE OPERATED Helsinki – Tallinn (Estonia) (3 hrs 30 mins; *(1,2)*; up to 4 per day).

1	NORDLANDIA	21473t	81	21.0k	153.4m	2048P	530C	40T	BA	FI
2	TRANSLANDIA	13700t	76	17.0k	135.8m	100P	280C	115T	A	FI

NORDLANDIA Built by AG Weser Seebeckswerft, Bremerhaven, Germany as the OLAU HOLLANDIA for *Olau Line* of Germany for the service between Vlissingen (Netherlands) and Sheerness (England). In 1989 she was replaced by a new vessel of the same name and she was sold to *Nordström & Thulin*. She was renamed the NORD GOTLANDIA and introduced onto *Gotlandslinjen* services between Gotland and the Swedish mainland. In 1997 she was purchased by *Eckerö Linjen* of Åland for delivery in early 1998, following the ending of *Nordström & Thulin's* concession to operate the Gotland services. She was renamed the NORDLANDIA and placed on the *Eckerö Line* Helsinki – Tallinn service, operating day trips. From May 2004 on some days to operate two round trips per day.

TRANSLANDIA Built by J J Sietas Werft, Hamburg, Germany as the TRANSGERMANIA for *Poseidon Schiffahrt OHG* of Germany interests for *Finncarriers-Poseidon* services between Finland and West Germany. In 1991 chartered to *Norse Irish Ferries* and used on their freight service between Liverpool and Belfast. In 1992 she was returned to *Finncarriers* and in 1993 sold to Cypriot interests for use in the Mediterranean and renamed the ROSEBAY. In 1994 chartered to *Stena Line* to inaugurate a new service between Harwich and Rotterdam (Frisohaven). In 1995 the service was switched to Hoek van Holland following the construction of a new linkspan. She also, during the summer, carried cars towing caravans, motor caravans and their passengers. In 1997 she was chartered to *Sally Freight* and renamed the EUROSTAR, operating between Ramsgate and Ostend. Later in 1997 she was renamed the EUROCRUISER. In 1998 she returned on charter to *Stena Line* and resumed the name ROSEBAY. In 1999 she was temporarily transferred to the Irish Sea. In autumn 2000 she was transferred to the Killingholme – Hoek van Holland service but was withdrawn

Color Fantasy *(Miran Hamidulla)*

Peter Wessel *(FERRYInformation)*

in 2001 when the delivery of the new STENA HOLLANDICA enabled the STENA SEARIDER to replace her. She was then sold to *Rederi AB Engship* of Sweden, renamed the TRANSPARADEN and chartered to *Botnia Link*. In 2002 chartered to *DFDS Tor Line* to operate between Kiel and Riga. In January 2003 transferred to the *Latlines* Lübeck – Riga route. Later in 2003 chartered to *SCF St Petersburg Line*. In May 2004 sold to *Eckerö Line* and operated between Helsinki and Tallinn and renamed the TRANSLANDIA.

ECKERÖ LINJEN

THE COMPANY *Eckerö Linjen* is an Åland Islands company.

MANAGEMENT Managing Director Björn Blomquist, **Marketing Director** Christer Lindman.

ADDRESS Torggatan 2, Box 158, AX-22100 Mariehamn, Åland.

TELEPHONE Administration +358 (0)18 28000, **Reservations** +358 (0)18 28300.

FAX Administration & Reservations +46(0)175 30820.

INTERNET Website www.eckerolinjen.fi *(Swedish, Finnish, English, German)*

ROUTE OPERATED Eckerö (Åland) – Grisslehamn (Sweden) (2 hrs; *(1,3)*; 5 per day).

1	ALANDIA	6754t	72	17.0k	108.7m	1320P	225C	32T	BA	AX
2	ECKERÖ	12131t	79	19.5k	121.0m	1500P	262C	26T	BA	AX
3	ROSLAGEN	6652t	72	18.7k	109.3m	1320P	225C	32T	BA	AX

ALANDIA Built by Jos L Meyer Werft, Papenburg, Germany as the DIANA for *Rederi AB Slite* of Sweden for *Viking Line* services. In 1979 she was sold to *Oy Vaasa-Umeå Ab* of Finland, renamed the BOTNIA EXPRESS and operated between Vaasa and Umeå. In 1982 she was sold to *Sally Line* of Finland; later that year she was sold to *Suomen Yritysrahoitus Oy* and chartered back. 1989 she was chartered to *Jacob Lines* for the route Skellefteå – Jakobstad under the marketing name 'Polar Princess' and in March 1990 she was placed on Helsinki-Tallinn cruises, but due to some customs/visa problems she never called at Tallinn. In 1992 she was sold to *Eckerö Linjen* and renamed the ALANDIA. She has also been used by subsidiary company *Eckerö Line*. In 2004 and also 2005 chartered to *RG-Line* to operate between Vaasa (Finland) and Umeå (Sweden) after the grounding of the CASINO EXPRESS.

ECKERÖ Built by Aalborg Værft A/S, Aalborg, Denmark as the JENS KOFOED for *BornholmsTrafikken*. Used on the Rønne – Copenhagen, Rønne – Ystad and (until December 2002) Rønne – Sassnitz services. Rønne – Copenhagen service became Rønne – Køge in September 2004. In October 2004 sold to *Eckerö Line* for delivery in May 2005. To be renamed and substantially rebuilt before entering service in late 2005 or early 2006. Now renamed ECKERÖ

ROSLAGEN Built by Jos L Meyer Werft, Papenburg, Germany as the VIKING 3 for *Rederi AB Sally* and used on *Viking Line* Baltic services. In 1976 she was sold to *Vaasanlaivat* of Finland for their service between Vaasa (Finland) and Umeå/Sundsvall (Sweden) and renamed the WASA EXPRESS. In 1982 *Vaasanlaivat* was taken over by *Rederi AB Sally* and in April 1983 she resumed the name VIKING 3, was transferred to *Sally Line* and used on the Ramsgate – Dunkerque service. She remained in the Channel during winter 1983/4 on freight-only services. However, in early 1984 she returned to *Vaasanlaivat* and resumed the name WASA EXPRESS. In 1988 she was sold to *Eckerö Linjen* and renamed the ROSLAGEN. During winter 1992/3 she operated between Helsinki and Tallinn for *Estonia New Line* and returned to *Eckerö Linjen* in the spring.

AG EMS

THE COMPANY *AG Ems* is a German public sector company.

MANAGEMENT Managing Director & Chief Executive B W Brons, **Marine Superintendent** J Alberts, **Marketing Manager & Assistant Manager** P Eesmann.

ADDRESS Am Aussenhafen, Postfach 1154, 26691 Emden, Germany.

TELEPHONE Administration & Reservations +49 (0)4921 8907-400 or +49 (0)4921 8907-406.

FAX Administration & Reservations +49 (0)4921 8907-405.

INTERNET Email info@ag-ems.de **Website** www.ag-ems.de *(German, Dutch)*

ROUTES OPERATED Conventional Ferries Emden (Germany) – Borkum (German Frisian Islands) (2 hrs; *(1,3,6)*; up to 4 per day), Eemshaven (Netherlands) – Borkum (55 mins; *(1,3,6)*; up to 4 per day). **Fast Ferries** Emden – Borkum (1 hr; *(2,4)*; up to 4 per day), Eemshaven – Borkum (30 mins; *(2,4)*; 1 per week in summer).

1	MÜNSTERLAND	1859t	86	15.5k	78.7m	1200P	70C	10L	BA	GY
2p»	NORDLICHT	435t	89	33.0k	38.8m	272P	0C	0L	-	GY
3	OSTFRIESLAND	1859t	85	15.5k	78.7m	1200P	70C	10L	BA	GY
4p»	POLARSTERN	636t	00	40.0k	45.0m	405P	0C	0L	-	GY
5p	WAPPEN VON BORKUM	287t	76	11.5k	42.8m	358P	0C	0L	-	GY
6	WESTFALEN	1812t	72	15.5k	77.9m	1200P	65C	10L	BA	GY

MÜNSTERLAND, OSTFRIESLAND Built by Martin Jansen GmbH & Co KG Schiffswerft, Leer, Germany for *AG Ems*.

NORDLICHT Fjellstrand 38m passenger only catamaran built Mandal, Norway for *AG Ems*.

POLARSTERN Oceanfast Ferries (Australia) 45m passenger only catamaran built at Henderson, Australia for another operator as the CARAIBE JET. This order was cancelled before delivery and she was sold to *AG Ems* after a period of lay-up, arriving in 2001.

WAPPEN VON BORKUM Built by Schiffswerft Schlömer GmbH & Co KG, Oldersum, Germany as the HANNOVER for *Friesland Fahrlinie* of Germany. In 1979 sold to *AG Ems* and renamed the STADT BORKUM. In 1988 sold to *ST-Line* of Finland, operating day trips from Rauma and renamed the PRINCESS ISABELLA. In 1994 returned to *AG Ems* and renamed the WAPPEN VON BORKUM.

WESTFALEN Built by C Cassens Schiffswerft, Emden, Germany for *AG Ems*. Rebuilt in 1994.

FINNLINES

THE COMPANIES *Finnlines plc* is a Finnish private sector company.

MANAGEMENT President Asser Ahleskog, **Vice-President** Simo Airas.

ADDRESS PO Box 197, Salmisaarenkatu 1, 00180 Helsinki, Finland. ***Sales and marketing of Finnlines Passenger Services*** Nordic Ferry Center Oy, Lönnrotinkatu 21, 00120 Helsinki, Finland.

TELEPHONE Administration +358 (0)10 34350. **Reservations *(Nordic Ferry Center Oy)*** +358 (0)9-2510 200.

FAX Administration +358 (0)10 3435200, **Reservations** +358 (0)9-2510 2022.

INTERNET *Finnlines* Email info@finnlines.fi ***Nordic Ferry Center*** info@ferrycenter.fi

Website *Finnlines* www.finnlines.fi *(English, Finnish, German)*

Nordic Ferry Center www.ferrycenter.fi/finnlines/en/index.shtml *(English)*

ROUTES OPERATED All year Helsinki – Travemünde (33 hrs; *(1,2,3,4)*; 1/2 per day) **Note** frequencies refer to services which convey passengers.

1	FINNHANSA	32531t	94	21.3k	183.0m	90P	-	236T	A2	FI
2	FINNPARTNER	32534t	94	21.3k	183.0m	90P	-	236T	A2	FI
3	FINNTRADER	32534t	95	21.3k	183.0m	90P	-	236T	A2	FI
4	TRANSEUROPA	32534t	95	21.3k	183.0m	90P	-	236T	A2	GY

FINNHANSA, FINNPARTNER, FINNTRADER 'Ro-pax' vessels built by Stocznia Gdanska SA, Gdansk, Poland for *Finnlines Oy* of Finland to provide a daily service conveying both freight and a limited number of cars and passengers on a previously freight-only route.

TRANSEUROPA 'Ro-pax' vessel built by Stocznia Gdanska SA, Gdansk, Poland for *Poseidon Schiffahrt* of Germany to operate on a joint service between Lübeck and Helsinki. In 1997 *Poseidon Schiffahrt* was acquired by *Finnlines* and in 2001 renamed *Finnlines Deutschland AG*.

Under Construction

5	NEWBUILDING 1	42000t	05	25.0k	216.0m	500P	-	300T	A2	FI
6	NEWBUILDING 2	42000t	06	25.0k	216.0m	500P	-	300T	A2	FI
7	NEWBUILDING 3	42000t	06	25.0k	216.0m	500P	-	300T	A2	FI

NEWBUILDING 1, NEWBUILDING 2, NEWBUILDING 3 Under construction by Fincantieri-Cantieri Nav Italiani SpA, Ancona, Italy to operate between Helsinki and Travemünde.

FINNLINK

THE COMPANY is a subsidiary of *Finnlines*.

MANAGEMENT Managing Director Christer Backman.

ADDRESS Satamatie 11, 21100, Naantali, Finland.

TELEPHONE Administration & Reservations +358 (0)10 436 7620.

FAX Administration & Reservations 358 (0)10 436 7660.

INTERNET Email finnlink@finnlink.fi **Website** www.finnlink.fi *(English, Finnish, Swedish)*

ROUTE OPERATED Naantali (Finland) – Kapellskär (Sweden) (6 hrs; *(8,9,10)*; 3 per day).

1	FINNARROW	25996t	96	21.0k	168.0m	200P	800C	154T	BA2	SW
2	FINNEAGLE	30500t	99	22.0k	188.3m	440P	-	176T	BA2	FI
3	FINNFELLOW	33796t	00	22.0k	188.3m	452P	-	216T	BA	FI

FINNARROW Built by Pt Dok Kodja Bahri, Kodja, Indonesia as the GOTLAND for *Rederi AB Gotland* for charter. In 1997 briefly chartered to *Tor Line* and then to *Nordic Trucker Line*, to operate between Oxelösund and St Petersburg (a ro-ro service). In June 1997 she was chartered to *SeaWind Line*, enabling a twice daily passenger service to be operated. In late 1997 she was sold to *Finnlines* and renamed the FINNARROW. She started operating twice weekly between Helsinki and Travemünde. During summer 1998 she was transferred to *FinnLink*; a bow door was fitted and she was modified to allow for two level loading. In 2003 transferred to *Nordö Link*. In 2005 returned to *FinnLink*.

FINNEAGLE 'Ro-pax' vessel built by Astilleros Españoles, Cadiz, Spain. Ordered by *Stena Ro-Ro* of Sweden and launched as the STENA SEAPACER 2. In 1998 sold, before delivery, to *Finnlines* and renamed the FINNEAGLE. Although expected to join her sister on the Helsinki – Travemünde route, on delivery in late 1999 she entered service with *FinnLink*. During winter 1999/2000 she was modified for two deck loading.

FINNFELLOW Ro-pax ferry built as the STENA BRITANNICA by Astilleros Españoles, Cadiz Spain for *Stena RoRo* and chartered to *Stena Line bv* to operate between Hoek van Holland and Harwich. In 2003 replaced by a new STENA BRITANNICA, sold to *Finnlines*, renamed the FINNFELLOW and placed on the Helsinki – Travemünde route. In 2004 transferred to *FinnLink*.

Under Construction

5	NEWBUILDING 4	42000t	07	25.0k	216.0m	500P	-	300T	A2	FI
6	NEWBUILDING 5	42000t	07	25.0k	216.0m	500P	-	300T	A2	FI

NEWBUILDING 4, NEWBUILDING 5 Under construction by Fincantieri-Cantieri Nav Italiani SpA, Ancona, Italy for *Finnlines* to operate for *FinnLink* between Naantali and Kapellskär.

FJORD LINE

THE COMPANY THE COMPANY *Fjord Line* is a Norwegian company, 100% owned by *Bergen-Nordhordland Rutelag AS (BNR).*

MANAGEMENT Managing Director Ove Solem, **Marketing Manager** Nils Henrik Geitle.

ADDRESS Skoltegrunnskaien, PO Box 6020, 5020 Bergen, Norway.

TELEPHONE Administration +47 55 54 87 00, **Reservations** +47 55 54 88 00.

FAX Administration & Reservations +47 55 54 86 01.

INTERNET Email fjordline@fjordline.com **Website** www.fjordline.com *(Norwegian, Danish, English)*

ROUTE OPERATED Passenger Service Bergen – Egersund (Norway) – Hanstholm (Denmark) (15 hrs 30 mins; *(1)*; 3 per week), Egersund – Hanstholm (6 hrs 45 mins; *(1)*; 7 per week in summer). Also UK route – see Section 1. **Freight Service** (8 hrs; *(2)*; 1 per day).

1	FJORD NORWAY	31356t	86	21.0k	161.0m	1800P	600C	100T	BA	NO
2F	LYGRA	7012t	79	14.0k	113.4m	0P	-	54T	A	NO

FJORD NORWAY Built by Schichau Seebeckwerft AG, Bremerhaven, Germany as the PETER PAN for *TT-Line* for the service between Travemünde and Trelleborg. In 1992 sold to *TT Line* of Australia (no connection) for use on their service between Port Melbourne (Victoria) and Hobart (Tasmania) and renamed the SPIRIT OF TASMANIA. In 2002 sold to *Nordsjøferger K/S* of Norway and renamed the SPIR. After modification work she was renamed the FJORD NORWAY and chartered to *Fjord Line.*

LYGRA Built by Ankerløkken Verft Glommen A/S, Frederikstad, Norway as the CENTRO AMERICA for *Liniera Naviera Pan Atlantica SA* of Puerto Rico. In 1984 sold to *K/S Bergenske og Euro Trans I/S (K/S Seatrans)* of Norway and renamed the NORNEWS SERVICE. In 1992 sold to *K/S Eurotrans A/S* of the Bahamas. In 1994 renamed the TRANS FJORD, operating between Hamburg and Skien (Norway). In 1997 sold to *Fjord Line* and renamed the LYGRA. In 1994 sold to *Flores Holding AS* of Norway and chartered back.

HH-FERRIES

THE COMPANY *HH-Ferries* is a Danish/Swedish private sector company. In 2002 it was acquired by *Steneo AB* of Sweden, part of the *Stena Group.*

MANAGEMENT Managing Director Lars Meijer, **Marketing Manager** Jon Cavalli-Björkman.

ADDRESS Atlantgatan 2, 252 25 Helsingborg, Sweden.

TELEPHONE Administration +46 (0)42-26 80 00, **Reservations *Denmark*** +45 49 26 01 55, ***Sweden*** +46 (0)42-19 8000.

FAX Administration & Reservations *Denmark* +45 49 26 01 56, ***Sweden*** +46 (0)42-28 10 70.

INTERNET Email admin@hhferries.se **Website** www.hhferries.se *(Swedish, Danish)*

ROUTE OPERATED Helsingør – Helsingborg (20 mins; *(1,2)*; every 30 minutes).

1	GITTE 3	4296t	87	12.7k	95.0m	300P	170C	20T	BA	DK
2	MERCANDIA IV	4296t	89	13.0k	95.0m	420P	170C	18L	BA	DK
3	MERCANDIA VIII	4296t	87	13.0k	95.0m	420P	170C	18L	BA	DK

GITTE 3 Built by North East Shipbuilders Ltd, Sunderland, UK as the SUPERFLEX DELTA for *Vognmandsruten*. In 1990 this company was taken over by *DIFKO* and she was renamed the DIFKO STOREBÆLT. In 1998, following the opening of the Great Belt fixed link, the service ceased and she was laid up. In 1999 she was chartered to *Easy Line* renamed the GITTE 3 and operated between Gedser and Rostock. Laid up after August 1999 except for brief periods on charter to *HH-Ferries*. In 2003 sold to *HH-Ferries*.

MERCANDIA IV Built by North East Shipbuilders Ltd, Sunderland, UK as the SUPERFLEX NOVEMBER for *Vognmandsruten* of Denmark. In 1989 sold to *Mercandia* and renamed the MERCANDIA IV. In 1990 she began operating on their *Kattegatbroen* Juelsminde – Kalundborg service. In 1996 she was transferred to their *Sundbroen* Helsingør – Helsingborg service. In 1997 the service and vessel were leased to *HH-Ferries*. In 1999 she was purchased by *HH-Ferries*. She has been equipped to carry dangerous cargo.

MERCANDIA VIII Built by North East Shipbuilders Ltd, Sunderland, UK as the SUPERFLEX BRAVO for *Vognmandsruten* of Denmark and used on their services between Nyborg and Korsør and Copenhagen (Tuborg Havn) and Landskrona (Sweden). In 1991 she was chartered to *Scarlett Line* to operate on the Copenhagen and Landskrona route. In 1993 she was renamed the SVEA SCARLETT but later in the year the service ceased and she was laid up. In 1996 she was purchased by *Mercandia*, renamed the MERCANDIA VIII and placed on their *Sundbroen* Helsingør – Helsingborg service. In 1997 the service and vessel was leased to *HH-Ferries*. In 1999 she was purchased by *HH-Ferries*.

HURTIGRUTEN

SERVICE The *'Hurtigruten'* is the *'Norwegian Coastal Express Service'*. It is part cruise, part passenger ferry, part cargo line and part car ferry (although this is a fairly minor part of the operation). In recent years the service has been operated by a consortium of two operators – *Ofotens og Vesteraalen Dampskibsselskab* and *Troms Fylkes Dampskibsselskap*. Plans to merge in 2002 to form a new company called *Nord Norges Dampskipsselskap* were dropped.

ADDRESS ***Ofotens og Vesteraalen Dampskibsselskab*** Postboks 43, 8501 Narvik, Norway, ***Troms Fylkes Dampskibsselskap*** 9005 Tromsø, Norway.

TELEPHONE Administration ***Ofotens og Vesteraalen D/S*** +47 76 96 76 96, ***Troms Fylkes D/S*** +47 77 64 82 00, **Reservations** ***Norway*** 810 30 000, ***UK*** +44 (0)20 7371 4011.

FAX Administration & Reservations ***Ofotens og Vesteraalen D/S*** +47 76 96 76 11, ***Troms Fylkes D/S*** +47 77 64 82 40, ***Reservations (UK)*** +44 (0)20 7371 4070.

INTERNET Email booking@ovds.no booking@tfds.no **Website** www.hurtigruten.no *(English, Norwegian)*

ROUTE OPERATED Bergen – Kirkenes with many intermediate calls. Daily departures throughout the year. The round trip takes just under 11 days.

1	FINNMARKEN	14000t	02	18.0k	133.0m	647P	50C	OL	SC	NO
2	KONG HARALD	11204t	93	18.0k	121.8m	691P	50C	OL	SC	NO
3	LOFOTEN	2621t	64	16.0k	87.4m	410P	4C	OL	C	NO
4	MIDNATSOL	14000t	03	18.0k	135.7m	626P	50C	-	SC	NO
5	NARVIK	6257t	82	18.0k	108.6m	550P	40C	OL	SC	NO
6	NORDKAPP	11386t	96	18.0k	123.3m	691P	50C	OL	SC	NO
7	NORDLYS	11204t	94	18.0k	121.8m	691P	50C	OL	SC	NO
8	NORDNORGE	11384t	97	18.0k	123.3m	691P	50C	OL	SC	NO
9	NORDSTJERNEN	2194T	56	15.5k	80.6m	192P	0C	OL	-	NO
10	POLARLYS	11341t	96	18.0k	123.0m	691P	50C	OL	SC	NO
11	RICHARD WITH	11205t	93	18.0k	121.8m	691P	50C	OL	SC	NO
12	TROLLFJORD	14000t	02	18.0k	135.7m	626P	50C	OL	SC	NO
13	VESTERÅLEN	6261t	83	18.0k	108.6m	550P	40C	OL	SC	NO

FINNMARKEN Built by Kvaerner Kleven Skeppsvarv, Ulsteinvik, Norway for *Ofotens og Vesteraalen D/S* to replace the LOFOTEN.

KONG HARALD Built by Volkswerft, Stralsund, Germany for *Troms Fylkes D/S.*

LOFOTEN Built by A/S Aker Mekaniske Verksted, Oslo, Norway for *Vesteraalens D/S.* In 1988 she was sold to *Finnmark Fylkesrederi og Ruteselskap.* In 1996 she was sold to *Ofotens og Vesteraalen D/S.* In 2002 she was replaced by the FINNMARKEN but continued to operate cruises and she substituted for the NORDNORGE during winter 2002/3 when she went to South America to operate cruises. This was repeated during winter 2003/4, winter 2004/5 and will be repeated during winter 2005/6.

MIDNATSOL Built by Fosen Mekaniske Verksted, Rissa, Norway for *Troms Fylkes D/S.*

NARVIK Built by Trondhjems MV, Aker Trondelag, Trondheim, Norway for *Ofoten D/S.* Since 1987 owned by *Ofotens og Vesteraalen D/S.*

NORDKAPP Built by Kvaerner Kleven, Ulsteinvik, Norway for *Ofotens og Vesteraalen D/S.* During winter 2005/6 to operate cruises down the coast to Chile and to Antarctica.

NORDLYS Built by Volkswerft, Stralsund, Germany for *Troms Fylkes D/S.*

NORDNORGE Built by Kvaerner Kleven, Ulsteinvik, Norway for *Ofotens og Vesteraalen D/S.* During winter since 2002/3 has operated cruises down the coast to Chile and to Antarctica.

NORDSTJERNEN Built by Blohm & Voss AG, Hamburg, Germany for *Det Bergenske Damskipsselskab A/S* of Norway and placed into Hurtigruten service. In 1979 sold to *Troms Fylkes D/S.* In 1994 placed on the Tromsø – Svalbard route. Returned to Hurtigruten in 1995 and then back to the Tromsø – Svalbard route in 1996. During winter 2005/6 to be chartered to *Ofotens og Vesteraalen D/S* to operate on the Hurtigruten whilst the NORDCAPP and NORDNORGE are operating in South America.

POLARLYS Built by Ulstein Verft A/S, Ulsteinvik, Norway for *Troms Fylkes D/S.*

RICHARD WITH Built by Volkswerft, Stralsund, Norway for *Ofotens og Vesteraalen D/S.* In 2002 sold to Norwegian interests and chartered back.

TROLLFJORD Built by Fosen Mekaniske Verksted, Rissa, Norway for *Troms Fylkes D/S.*

VESTERÅLEN Built by Kaarbös Mekaniske Verksted A/S, Harstad, Norway for *Vesteraalens D/S.* Since 1987 owned by *Ofotens og Vesteraalen D/S.*

JETLINK

THE COMPANY *Jetlink AS* is a Norwegian private sector company.

MANAGEMENT Managing Director & Marketing Manager Captain Lars Kjellerod.

ADDRESS Postboks 282, 1502 Moss, Norway.

TELEPHONE Administration and Reservations +47 69 24 39 00

FAX Administration & Reservations +47 69 24 39 01.

INTERNET Email post@jetlink.no **Website** www.jetlink.no *(Norwegian)*

ROUTE OPERATED Moss (Norway) – Frederikshavn (Denmark) (4 hrs; *(1)*; 1 per day). Service due to start May 2005.

The start of this service is uncertain as the designated vessel has been sold to another operator.

KYSTLINK

THE COMPANY *KystLink AS* is a Norwegian private sector company.

MANAGEMENT Managing Director *not known*, **Marketing Manager** *not known*.

ADDRESS Kongshavn 8, NO-3970 Langesund, Norway.

TELEPHONE Administration & Reservations +47 35 96 68 00.

FAX Administration & Reservations +47 35 96 68 01.

INTERNET Email post@kystlink.no **Website** www.kystlink.com *(Norwegian, English)* www.kystlink.de *(German)*

ROUTE OPERATED Langesund – Hirtshals (6 hours; *(1)*; 1 per day).

1	ENVOY	18653t	79	18.2k	150.0m	107P	-	142T	A	BD

ENVOY Built by Mitsui Engineering and Shipbuilding Co Ltd, Tamano, Japan as the IBEX for *P&O* for *Pandoro* Irish sea services. In 1980 chartered to *North Sea Ferries*, renamed the NORSEA and used on the Ipswich – Rotterdam service. In 1986 she was renamed the NORSKY. In 1995 she returned to *Pandoro* and was re-registered in Bermuda. Later in 1995 she resumed her original name of IBEX. An additional deck was added in 1996. In late 1997 she was renamed the EUROPEAN ENVOY. In November 2001 transferred to the Mostyn – Dublin service. In 2004 the Mostyn route closed and she was sold to *Stena RoRo* who resold her to *Kystlink* who renamed her the ENVOY.

LISCO BALTIC SERVICE

THE COMPANY *AB Lisco Baltic Service* is a Lithuanian company, owned by *DFDS A/S of Denmark*; they purchased the company from the Lithuanian Government in 2001. Passenger and cargo services are marketed by *AB Lisco Baltic Service* and ticket sales in Lithuania by *Krantas Travel*. In 2004 the services of *Latlines* were taken over by *DFDS Tor Line*, who themselves operated a service between Germany and Latvia. Services are now operated by *Lisco Baltic Service*.

ADDRESS 24 J Janonio Str, Klaìpeda 92251, Lithuania.

TELEPHONE Administration *LISCO Baltic Service (Klaìpeda)* + 370 46 393600, ***Lübeck*** +49 (0)451 7099697, **Reservations *LISCO (Klaìpeda):*** + 370 46 393616, ***LISCO/Krantas Travel*** + 370 46 395050, ***DFDS Tor Line/LISCO Baltic Service (Karlshamn)*** + 46 (0)45433680, ***LISCO Baltic Service GmbH (Kiel)*** + 49 (0)431 2097 6444 – cargo, + 49 (0)43120976420 – passengers, ***Riga*** +371 7353523, ***Lübeck*** +49 (0)451 7099685.***Riga*, Reservations**

FAX Administration *LISCO Baltic Service (Klaìpeda)* + 370 46 393601, ***Riga*** +371 7353066, ***Lübeck*** +49 (0)451 7099687, **Reservations *LISCO (Klaìpeda)*** + 370 46 393606, ***LISCO/Krantas Travel*** + 370 46 395052, ***DFDS Tor Line/LISCO Baltic Service (Karlshamn)*** + 46 (0)45433689, ***LISCO Baltic, Service GmbH (Kiel)*** + 49 (0)431 2097 6555, ***Riga*** +371 7353071, ***Lübeck*** +49 (0)451 7099687.

INTERNET Email – Administration ml5@lisco.lt office.riga@dfdstorline.com **Reservations *LISCO (Klaìpeda)*** booking@lisco.lt ***LISCO/Krantas Travel*** Cargo.de@lisco.lt passengers@lisco.lt ***DFDS Tor Line/LISCO Baltic Service (Karlshamn)*** Karlshamn@dfdstorline.com, ***LISCO Baltic Service GmbH (Kiel)***: cargo@lisco-baltic-service.de passage@lisco-baltic-service.de passenger.riga@dfdstorline.com

Websites www.lisco.lt *(English, German)* www.lisco.se *(English, Swedish)* www.dfdstorline.lv *(Latvian, English, Russian)*

ROUTES OPERATED Klaìpeda (Lithuania) – Kiel (Germany) (21 hrs; *(3)*; 6 per week) (joint with *Scandlines Euroseabridge* under the 'Kiel-Klaìpeda-Express' name), Klaìpeda – Karlshamn (Sweden) (14 hrs *(1,4)*; 6 per week), Klaìpeda – Gdynia (Poland) (8 hrs *(1)*; 1 per week), Klaìpeda – Sassnitz (Germany) (18 hrs; *(2)*; 3 per week (joint with *Scandlines Euroseabridge*)), Lübeck (Germany) – Riga (Latvia) (32 hrs; *(5,7)*; 3 per week), Lübeck (Germany) – Ventspils (Latvia) (26 hrs; *(5)*; 1 per week).

1	KAUNAS	25606t	89	15.8k	190.9m	235P	460C	108Tr	A2	LT
2	KLAÌPEDA	21890t	87	15.8k	190.9m	12P	250C	110Tr	A	LT
3	LISCO GLORIA	20600t	02	22.0k	199.4m	308P	316C	180T	A2	LT
4	LISCO PATRIA	18332t	92	18.0k	153.6m	250P	-	132T	BA2	LT
5	MERMAID II	13730t	72	17.5k	137.3m	69P	170C	84T	AS	LV
6	PALANGA	11630t	79	19.0k	126.5m	130P	300C	70T	A2	LT
7	VILNIUS	22341t	87	15.8k	190.9m	132P	460C	108Tr	A2	LT

KAUNAS train ferry built by VEB Mathias-Thesen-Werft, Wismar, Germany (DDR) for *Lisco* of the former Soviet Union and used by to operate between Klaìpeda and Mukran in Germany (DDR). This was part of series of vessels built to link the USSR and Germany (DDR), avoiding Poland. In 1994 she was modified to increase passenger capacity in order to offer limited passenger facilities and placed on Klaìpeda – Kiel service. In 2003 transferred to the Klaìpeda – Karlshamn route. Early in 2004 chartered to *DFDS Tor Line* to operate between Lübeck and Riga. In 2005 transferred to the Klaìpeda – Karlshamn route.

KLAÌPEDA train ferry built by VEB Mathias-Thesen-Werft, Wismar, Germany (DDR) for *Lisco* of the former Soviet Union and used by to operate between Klaìpeda and Mukran in Germany (DDR). This was part of series of vessels built to link the USSR and Germany (DDR), avoiding Poland. Unlike other ships of this series, she was not converted to full ro-pax format. Operates on the Klaìpeda –Sassnitz route.

LISCO GLORIA Built by Stocznia Szczecinska im A Warskiego, Szczecin, Poland for *Lloyd Sardegna* of Italy as the GOLFO DEI CORALLI for operating between Italy and Sardinia. However, due to late delivery the order was cancelled. In 2002 purchased by *DFDS Tor Line*, renamed the DAN GLORIA and, in Autumn 2002 placed on the Esbjerg – Harwich service. In June 2003 replaced by modified sister vessel DANA SIRENA (see Section 1) and sold to *Lisco Baltic Service* and renamed the LISCO GLORIA. Operates between Klaìpeda and Kiel.

LISCO PATRIA Ro-pax vessel built by Fosen Mekaniske Verksted, Trondheim, Norway for *Stena RoRo*. After a short period with *Stena Line* on the Hoek van Holland – Harwich service, she was chartered to *Sealink Stena Line* for their Southampton – Cherbourg route, initially for 28 weeks. At the end of the 1992 summer season she was chartered to *TT-Line* to operate between Travemünde and Trelleborg and was renamed the TT-TRAVELLER. In late 1995, she returned to *Stena Line*, resumed the name STENA TRAVELLER and inaugurated a new service between Holyhead and Dublin. In autumn 1996 she was replaced by the STENA CHALLENGER (18523t, 1991). In early 1997 she was again chartered to *TT-Line* and renamed the TT-TRAVELLER. She operated on the Rostock – Trelleborg route. During winter 1999/2000, her passenger capacity was increased to 250 and passenger facilities renovated. In early 2002 the charter ended and she was renamed the STENA TRAVELLER, chartered to *Stena Line* and placed on their Karlskrona – Gdynia service. Charter ended May 2003. Sold to *Lisco Baltic Service* and renamed the LISCO PATRIA. Placed on the Klaìpeda – Karlshamn service.

MERMAID II Built by Oy Wärtsilä AB, Turku, Finland as the HANZ GUTZEIT and chartered to *Finncarriers* for Finland – Germany service. In 1982 she was sold to *EFFOA* of Finland and renamed the CAPELLA. She continued to be chartered to *Finncarriers* and this charter continued under a number of subsequent owners. In 1986 she was renamed the CAPELLA AV STOCKHOLM. In 1988 she was renamed the FINNMAID. In 1989 she was placed on the *FinnLink* service between Uusikaupunki (Finland) and Hargshamn (Sweden). In 1997, this service was transferred to the Kapellskär – Naantali route. In 1998 she was replaced by the FINNARROW and, after service on *Finncarriers'* Finland – Germany routes, was laid up. In 2000 she was chartered to *VV-Line* who operated between Vastervik in Sweden and Ventspils in Latvia. Passenger capacity was raised from 48 to 69. She was then owned by *Rederi AB Gustaf Erikson* of Åland but in 2001 purchased by *VV-Line* and renamed the MERMAID II. In 2003 chartered to *DFDS Tor Line* to operate between Kiel and Riga. In March service transferred to Lübeck. In 2003 sold to *Skandia Liv Ab* of Panama. In 2004 chartered to *LISCO* and placed on the Klaìpeda – Karlshamn route. In 2005 to operate once weekly between Lübeck and Riga and once weekly between Lübeck and Ventspils.

PALANGA Built by Ateliers & Chantiers du Havre, Le Havre, France; rebuilt at in 1992 to increase passengers capacity. In 1996 sold to *Lisco*, renamed the PALANGA; used on a Klaìpeda – Stockholm route (later Klaìpeda – Karlshamn). In 2003 briefly operated the Klaìpeda – Kiel route but returned to the Karlshamn route when the LISCO GLORIA entered service. In June 2004 chartered to *DFDS Tor Line* to operate between Harwich and Esbjerg. In August she inaugurated a Lübeck – Ventspils service for *DFDS Tor Line*. In December 2004 chartered to *Trasmediterránea* to operate between Spain and the Canary Islands.

VILNIUS train ferry as KAUNAS. Operated on the Klaìpeda – Kiel service until June 2003. Later chartered to *DFDS Tor Line* to operate between Lübeck and Riga.

AB Lisco Baltic Service also own the TOR NERINGA which is currently on charter to *DFDS Tor Line*. She is shown under *DFDS Tor Line* in Section 3.

MOLS-LINIEN

THE COMPANY *Mols-Linien A/S* is a Danish private sector company; previously a subsidiary of *J Lauritzen A/S*, it was, in 1988 sold to *DIFKO No LXII (Dansk Investeringsfond)*. Since 1994 shares in the company have been traded on the stock exchange. In January 1999 a 40% share in the company was acquired by *Scandlines Danmark A/S*. Their *Scandlines Cat-Link* Århus – Kalundborg service became part of *Mols-Linien* in February 1999 and the service was switched from Kalundborg to Odden in April 1999. The Ebeltoft – Odden ro-pax vessels were transferred to the Århus – Kalundborg route in January 2000.

MANAGEMENT Managing Director Preben Wolff, **Marketing Manager** Christian Hingelberg.

ADDRESS Færgehavnen, 8400 Ebeltoft, Denmark.

TELEPHONE Administration +45 89 52 52 00, **Reservations** +45 70 10 14 18.

FAX Administration +45 89 52 52 90, **Reservations** +45 89 52 52 92.

INTERNET Email administration@mols-linien.dk **Website** www.mols-linien.dk *(Danish)*

ROUTES OPERATED Ro-pax Ferries Århus (Jutland) – Kalundborg (Sealand) (2 hr 40 mins; *(4,6)*; 7 per day). **Fast Ferries** Århus – Odden (Sealand) (1 hr 5 mins; *(1,2)*; every 1 hrs 30 mins), Ebeltoft (Jutland) – Odden (45 mins; *(3,5)*; two hourly).

1»	MADS MOLS	5619t	98	43.0k	91.3m	800P	220C	-	A	DK
2»	MAX MOLS	5617t	98	43.0k	91.3m	800P	220C	-	A	DK
3»	MAI MOLS	3971t	96	43.4k	76.1m	450P	120C	-	BA	DK
4	MAREN MOLS	14221t	96	19.0k	136.4m	600P	344C	82L	BA2	DK
5»	MIE MOLS	3971t	96	43.4k	76.1m	450P	120C	-	BA	DK
6	METTE MOLS	14221t	96	19.0k	136.4m	600P	344C	82L	BA2	DK

MADS MOLS InCat 91 metre catamaran, built speculatively at Hobart, Tasmania, Australia. In spring 1998, following *InCat's* acquisition of a 50% share in *Scandlines Cat-Link A/S*, she was chartered by *Nordic Catamaran Ferries K/S* to that company, operating between Århus and Kalundborg and named the CAT-LINK V. She is the current holder of the Hales Trophy for fastest crossing of the Atlantic during her delivery voyage between the USA and Falmouth, UK. In 1999 the charter was transferred to *Mols-Linien*, she was renamed the MADS MOLS and operated between Århus and Odden. Charter to end in July 2005.

MAX MOLS InCat 91 metre catamaran, built speculatively at Hobart, Tasmania, Australia. In spring 1998, following *InCat's* acquisition of a 50% share in *Scandlines Cat-Link A/S*, She was sold to that company and named the CAT-LINK IV. In 1999 purchased by *Mols-Linien* and renamed the MAX MOLS. In 2000 chartered to *Marine Atlantic* of Canada to operate between Port aux Basques (Newfoundland) and North Sydney (Nova Scotia). Returned to *Mols-Linien* in autumn 2000. In summer 2002 chartered to *Riga Sea Lines* to operate between Riga and Nynäshamn. Returned to *Mols-Linien* in autumn 2002. In 2004 chartered to *P&O Ferries* to operate between Portsmouth and Caen. Operated under the marketing name, the 'Caen Express'. In November 2004 returned to *Mols-Linien* and placed on the Århus – Odden route to enhance the service as second vessel.

Mercandia IV *(Miles Cowsill)*

Lisco Gloria *(FotoFlite)*

MAI MOLS Danyard SeaJet 250 catamaran built by Danyard-Aalborgs A/S, Aalborg, Denmark for *Mols-Linien*.

MAREN MOLS, METTE MOLS Ro-pax vessels built by Ørskovs Christensens Staalskibsværft A/S, Frederikshavn, Denmark for *Mols-Linien*. Initially operated on the Ebeltoft – Odden route. In January 2000 switched to the Århus – Kalundborg route.

MIE MOLS Danyard SeaJet 250 catamaran built at Danyard-Aalborgs A/S, Aalborg, Denmark for *Mols-Linien*.

REEDEREI NORDEN-FRISIA

THE COMPANY *Aktiengesellschaft Reederei Norden-Frisia* is a German public sector company.

MANAGEMENT President/CEO Mr C U Stegmann, **Managing Director/CFO** Prok. Graw, **Technical Manager** Prok. H Stolle.

ADDRESS Postfach 1262, 26534 Norderney, Germany.

TELEPHONE Administration +49 (0)4932 9130.

FAX Administration +49 (0)4932 9131320.

INTERNET Email info@reederei-frisia.de **Website** www.reederei-frisia.de *(German)*

ROUTES OPERATED ***Car Ferries & Passenger Ferries*** Norddeich (Germany) – Norderney (German Frisian Islands) (1 hr; *(2,4,5)*; up to 15 per day), Norddeich – Juist (German Frisian Islands) (1 hr 20 mins; *(3,6)*; up to 15 per day), ***Passenger only fast ferry*** Norderney – Helgoland (1 hr 30 mins; *(1)*), Langeoog – Helgoland (1 hr 15 min; *(1)*), Wangerooge – Helgoland (1 hr 15 min; *(1)*), Hooksiel-Außenhafen – Sylt (2 hrs 30 mins; *(1)*), Hooksiel-Außenhafen – Amrum/Föhr (2 hrs; *(1)*), Wilhelmshaven – Helgoland (1 hr 20 mins; *(1)*).

1p»	CAT NO 1	963t	99	40.0k	52.6m	432P	-	-	-	GY
2	FRISIA I	1020t	70	12.3k	63.7m	1500P	53C	-	BA	GY
3	FRISIA II	1125t	78	12.0k	63.3m	1340P	53C	-	BA	GY
4	FRISIA IV	1600t	02	12.0k	71.7m	1342P	60C	-	BA	GY
5	FRISIA V	1007t	65	11.0k	63.8m	1442P	53C	-	BA	GY
6	FRISIA VI	768t	68	12.0k	54.9m	1096P	35C	-	BA	GY
7F	FRISIA VII	363t	84	12.0k	53.0m	12P	30C	-	BA	GY
8p	FRISIA IX	571t	80	11.0k	57.0m	785P	-	-	-	GY
9p	FRISIA X	187t	72	12.0k	36.3m	290P	-	-	-	GY

CAT NO I Built by Fremantle, Western Australia for *Reederei Norden-Frisia* to inaugurate seasonal daily services to Helgoland and other North Sea Islands.

FRISIA I, FRISIA II, FRISIA V, FRISIA VI Built by Jos L Meyer Werft, Papenburg, Germany for *Reederei Norden-Frisia*. Passenger capacities relate to the summer seasons. Capacity is reduced during the winter.

FRISIA IV Built by Schiffswerft und Maschinenfabrik Cassens GmbH, Emden, Germany for *Reederei Norden-Frisia* to replace the FRISIA VIII.

FRISIA VII Built by Schlömer Werft, Oldersum, Germany for *Reederei Norden-Frisia*. Conveys ro-ro freight to Norderney and Juist.

FRISIA IX, FRISIA X Built by Schiffswerft Julius Diedrich GmbH & Co. KG, Oldersum, Germany for *Reederei Norden-Frisia*. The FRISIA IX was built to convey 9 cars at the bow end but is now used mainly in passenger only mode. These ships are generally used for excursions.

RØMØ-SYLT LINIE

THE COMPANY *Römö-Sylt Linie GmbH* is a German company, a subsidiary of *FRS (Förde Reederei Seetouristik)* of Flensburg. (Note: although, being a Germany company, the company's official name uses the German version of the umlaut (ö), the company trades in both Denmark and Germany using the Danish version (ø).

MANAGEMENT Managing Director P Rathke.

ADDRESS ***Germany*** Am Fähranleger, 25992 List, Germany, ***Denmark*** Kilebryggen, 6792 Rømø, Denmark.

TELEPHONE Administration (Germany) +49 (0)4651 870475, **Reservations (Denmark)** +45 73 75 53 03.

FAX Administration (Germany) +49 (0)4651 871446, **Reservations (Denmark)** +45 73 75 53 05.

INTERNET Email romo-sylt@post12.tele.dk **Website** www.romo-sylt.dk *(Danish, German)*

ROUTE OPERATED List (Sylt, Germany) – Havneby (Rømø, Denmark) (45 mins; *(1,2)*; variable – hourly at peaks). **Note** the island of Rømø is linked to the Danish mainland by a road causeway; the island of Sylt is linked to the German mainland by a rail-only causeway on which cars are conveyed on shuttle wagons.

1	VIKINGLAND	1963t	74	11.0k	68.3m	420P	60C	8L	BA	GY
2	WESTERLAND	1509t	71	11.0k	58.3m	400P	40C	5L	BA	GY

VIKINGLAND, WESTERLAND Built by Husumer Schiffswerft, Husum, Germany for *Rømø-Sylt Linie*.

Under Construction

3	NEWBUILDING	-	05	16.0k	88.2m	600P	80C	10L	BA	GY

NEWBUILDING Under construction by Fiskerstrand Verft A/S, Aalesund, Norway for *Rømø-Sylt Linie*.

SAAREMAA LAEVAKOMPANII

THE COMPANY *Saaremaa Laevakompanii* is an Estonian company, founded in 1992.

MANAGEMENT General Director Tõnis Rihvk.

ADDRESS AS Kohtu, 93812 Kuressaare, Estonia.

TELEPHONE Administration +372-45-24350, **Reservations** +372-45-24444.

FAX Administration +372-45-24355, **Reservations** +372-45-24373.

INTERNET Email slk@laevakompanii.ee **Website** www.laevakompanii.ee *(Estonian, English)*

ROUTE OPERATED Kuivastu – Virtsu (30 mins; *(3,6,8,10)*; up to 12 per day), Heltermaa (Hiiumaa) – Rohuküla (1 hr 30 mins; *(2,4,5)*; 5 per day), Rohuküla – Sviby (45 mins; *(1 4)*; up to 3 per day), Triigi – Sõru (1 hr 5 mins; *(9)*; up to 2 per day). **From May 2005** Mõntu (Saaremaa) – Ventspils (Latvia) ((7)).

1	HARILAID	1028t	85	9.9k	49.9m	120P	35C	0L	BA	ES
2	HIIUMAA	1549t	66	10.0k	53.8m	175P	28C	0L	BA	ES
3	KOGUVA	1305t	79	10.6k	55.5m	204P	41C	0L	BA	ES
4	KÖRGELAID	1028t	87	9.9k	49.9m	200P	35C	0L	BA	ES
5	OFELIA	3638t	68	14.4k	74.4m	600P	110C	12L	BA	ES
6	REGULA	1157t	71	14.5k	71.2m	580P	105C	12L	BA	ES
7	SCANIA	3474t	72	14.5k	74.2m	400P	80C	12L	BA	ES
8	ST OLA	4833t	71	16.0k	85.9m	500P	140C	12L	BA	ES
9	VARDO	116t	62	9.0k	27.1m	47P	17C	0L	BA	ES
10	VIIRE	4101t	88	12.7k	95.8m	469P	140C	36L	BA	ES

Wawel *(Jakub Bogucki/Green Ship)*

Skåne *(John May)*

HARILAID Built by Riga Shiprepair Yard, Riga, Latvia (USSR) for *ESCO* of Estonia. In 1994 transferred to *Saaremaa Laevakompanii.*

HIIUMAA Built by P Høivolds Mekaniska Verksted A/S, Kristiansand, Norway as the TAARS for *Sydfyenske Dampskibsselskab A/S* of Denmark. Initially used on the Spodsbjerg – Nakskov route. In 1975 transferred to the Spodsbjerg – Taars route. Withdrawn in 1984 and in 1985 sold to *ESCO* of Estonia and renamed the HIIUMAA; she was placed on the Heltermaa – Rohuküla route. In 1992 renamed the HIIUMAA 2. In 1994 transferred to *Saaremaa Laevakompanii.* In 1996 renamed the HIIUMAA.

KOGUVA Built by Riga Shiprepair Yard, Riga, Latvia (USSR) for *ESCO* of Estonia. In 1994 transferred to *Saaremaa Laevakompanii.*

KÖRGELAID Built by Riga Shiprepair Yard, Riga, Latvia (USSR) for *ESCO* of Estonia. In 1994 transferred to *Saaremaa Laevakompanii.*

OFELIA Built by Krögerwerft GmbH, Rendsburg, Germany for *Svenska Rederi-AB Öresund* of Sweden and used on the Limhamn – Dragør service. In 1980 sold to *Scandinavian Ferry Lines.* In 1990 transferred to *SweFerry* and later to *Scandlines AB.* In 1997 sold to *Saaremaa Laevakompanii.*

REGULA Built by Jos L Meyer, Papenburg, Germany for *Stockholms Rederi AB Svea* of Sweden for the service between Helsingborg and Helsingør operated by *Linjebuss International AB* (a subsidiary company). In 1980 she was sold to *Scandinavian Ferry Lines.* During winter 1984/85 she was rebuilt to increase vehicle and passenger capacity. In 1991 ownership was transferred to *SweFerry* and operations to *ScandLines* on the Helsingborg – Helsingør service. Ownership later transferred to *Scandlines AB.* In 1997 sold to *Saaremaa Laevakompanii.*

SCANIA Built by Aalborg Værft A/S, Aalborg, Denmark for *Svenska Rederi-AB Öresund* of Sweden and used on the Limhamn – Dragør service. In 1980 sold to *Scandinavian Ferry Lines.* In 1990 transferred to *SweFerry* and later to *Scandlines AB.* In 1999 sold to *Saaremaa Laevakompanii.* In 2005 inaugurated a new service to Ventspils, Latvia.

ST OLA Built by Jos L Meyer, Papenburg Germany as the SVEA SCARLETT for *Stockholms Rederi AB Svea* of Sweden and used on the *SL (Skandinavisk Linjetrafik)* service between Copenhagen (Tuborg Havn) and Landskrona (Sweden). In 1980 she was sold to *Scandinavian Ferry Lines* of Sweden and *Dampskibsselskabet Øresund A/S* of Denmark (jointly owned). Initially she continued to serve Landskrona but later that year the Swedish terminal became Malmö. In 1981 she operated on the Helsingborg – Helsingør service for a short while, after which she was withdrawn and laid up. In 1982 she was sold to *Eckerö Linjen* of Finland, renamed the ECKERÖ and used on services between Grisslehamn (Sweden) and Eckerö (Åland Islands). In 1991 she was sold to *P&O Scottish Ferries* and renamed the ST OLA. In March 1992 she replaced the previous ST OLA (1345t, 1974) on the Scrabster – Stromness service. In September 2002 withdrawn and sold to *Saaremaa Laevakompanii.* Used on the Kuivastu – Virtsu route.

VARDO Built by Rauma-Repola Oy, Rauma, Finland as the VÅRDÖ for *Ålands Landskapsstyrelse* of Åland, Finland. In 1997 sold to *Saaremaa Laevakompanii* and renamed the VARDO.

VIIRE Built as the SUPERFLEX CHARLIE by North East Shipbuilders Ltd, Sunderland, UK for *Vognmandsruten* of Denmark to establish a new service between Korsør (Fyn) and Nyborg (Sjælland). In 1990 this company was taken over by DIFKO and she was renamed the DIFKO KORSØR. In 1998, following the opening of the Great Belt road fixed link, the service ceased and she was laid up. In 1999 chartered to *Saaremaa Laevakompanii* and renamed the VIIRE. In 2000, the charter ended and she returned to lay-up in Denmark. In 2001 she was again chartered to *Saaremaa Laevakompanii.* Used on the Kuivastu – Virtsu route.

SCANDLINES (DENMARK & GERMANY)

THE COMPANY *Scandlines AG* is a German company, 50% owned by *DB Cargo AG*, a subsidiary of *Deutsche Bahn AG (German Railways)* (which is owned by the Federal Government) and 50% owned by the Kingdom of Denmark. In 1998 it took over *DFO (Deutsche Fährgesellschaft Ostsee mbH)* of Germany (renamed *Scandlines Deutschland GmbH*) and *Scandlines A/S* of Denmark (renamed *Scandlines Danmark A/S*). A 50% share in *Euroseabridge GmbH* was acquired in 1998 (by *Scandlines A/S* of Denmark before the merger with *DFO*) and the remaining 50% in 1999; the name was changed to *Scandlines Euroseabridge GmbH*.

Scandlines A/S was formerly *DSB Rederi A/S* and before that the Ferries Division of *DSB (Danish State Railways)*. *DFO* was formed in 1993 by the merging of the Ferries Divisions of *Deutsche Bundesbahn (German Federal Railways)* (which operated in the Federal Republic of Germany) and *Deutsche Reichsbahn (German State Railways)* (which operated in the former DDR).

Stena Line owned *Scandlines AB* of Sweden also trades under this name but remains a separate company. Danish domestic routes are operated by subsidiary company *Scandlines Sydfynske A/S*, and are marketed as part of the *Scandlines* network.

MANAGEMENT Managing Director Peter Grønlund, **Finance Director** Uwe Bakosch.

ADDRESS ***Denmark*** Dampfærgevej 10, 2100 Copenhagen Ø, Denmark. ***Germany*** Hochhaus am Fährhafen, 18119 Rostock-Warnemünde, Germany.

TELEPHONE Administration ***Denmark*** +45 33 15 15 15, ***Germany*** +49 (0)381 5435-0, **Reservations** ***Denmark*** +45 33 15 15 15, ***Germany*** +49 (0)1805-7226 354637.

FAX Administration ***Denmark***, ***Germany*** +49 (0) 381 54 35 678

INTERNET Email scandlines@scandlines.dk **Websites** www.scandlines.dk *(Danish)* www.scandlines.com *(German, English)*

ROUTES OPERATED

Helsingør (Sealand, Denmark) – Helsingborg (Sweden) (25 mins; *(6,23)*; every 20 mins) (joint with *Scandlines AB* of Sweden), Rødby (Lolland, Denmark) – Puttgarden (Germany) (45 mins; *(2,7,14,15,17 (7 road freight only))*; half hourly train/vehicle ferry + additional road freight only sailings), Gedser (Falster, Denmark) – Rostock (Germany) (2 hrs; *(8,13)*; every 2 hours (less frequent Dec – Feb), Rostock (Germany) – Trelleborg (Sweden) (5 hrs 45 mins (7 hrs night); *(9)*; 3 per day) (joint with *Scandlines AB* of Sweden), Sassnitz (Germany) – Trelleborg (3 hrs 45 mins; *(16)*; 4-5 per day) (joint with *Scandlines AB* of Sweden).

Danish domestic services operated by subsidiary *Scandlines Sydfynske A/S (formerly Sydfyenske Dampskibsselskab (SFDS))* and forming part of ***Scandlines*** network, Fynshav (Als) – Bøjden (Fyn) (50 mins; *(22)*; 6-8 per day), Esbjerg (Jutland) – Nordby (Fanø) (12 mins; *(4,10,19)*; every 20-40 mins), Spodsbjerg (Langeland) – Tårs (Lolland) (45 mins; *(5,11,20)*; hourly).

Germany – Latvia & Lithuania routes operated by subsidiary *Scandlines Euroseabridge GmbH* Kiel – Klaìpeda (Lithuania) (20 hrs; *(21, Lisco vessel)*; 6 per week (joint with *Lisco Baltic Service* of Lithuania under the *Kiel-Klaìpeda-Express* name)), Sassnitz (Mukran) (Germany) – Klaìpeda (Lithuania) (18 hrs; *(Lisco vessel)*; 3 per week (joint with *Lisco Baltic Service* of Lithuania)), Rostock (Germany) – Liepaja (Latvia) (24 hrs; *(24)*; 2 per week).

Denmark – Lithuania freight route operated by *Scandlines Balticum Seaways division* Århus (Denmark) – Aabenraa (Denmark) – Klaìpeda (Lithuania) (from 30 hrs; *(18)*; 2 per week from Aabenraa (1 westbound via Århus), 1 per week from Århus via Aabenraa).

Sweden – Latvia routes operated by *Scandlines Amber Line* Karlshamn (Sweden) – Liepaja (Latvia) (17 hrs; *(12)*; 3 per week), Nynäshamn (Sweden) – Ventspils (Latvia) (12 hrs; *(3)*; 5 per week).

1	ASK	13294t	82	18.0k	171.0m	186P	-	104T	AS	DK
2	DEUTSCHLAND	15187t	97	19.0k	142.0m	900P	294C	30Lr	BA2	GY

3	FELLOW	14297t	73	18.0k	137.3m	64P	170C	84T	AS	FI
4	FENJA	751t	98	11.5k	49.9m	396P	34C	4L	BA	DK
5	FRIGG SYDFYEN	1676t	84	12.5k	70.1m	338P	50C	8L	BA	DK
6	HAMLET	10067t	97	14.0k	111.2m	1000P	244C	34L	BA	DK
7F	HOLGER DANSKE	2779t	76	14.5k	86.8m	12P	55C	12L	BA	DK
8	KRONPRINS FREDERIK	16071t	81	19.5k	152.0m	1082P	210C	46T	BA	DK
9	MECKLENBURG-VORPOMMERN	36185t	96	20.0k	199.9m	600P	90C	230Tr	A2	GY
10	MENJA	751t	98	11.5k	49.9m	396P	34C	4L	BA	DK
11	ODIN SYDFYEN	1698t	82	12.5k	70.4m	338P	50C	8L	BA	DK
12	PETERSBURG	25353t	86	16k	190.8m	144P	329C	110T	A2	LB
13	PRINS JOACHIM	16071t	80	19.5k	152.0m	1080P	-	46Lr	BA	DK
14	PRINS RICHARD	14621t	97	19.0k	142.0m	900P	294C	36Lr	BA2	DK
15	PRINSESSE BENEDIKTE	14621t	97	19.0k	142.0m	900P	294C	36Lr	BA2	DK
16	SASSNITZ	21154t	89	19.0k	171.5m	1000P	390C	50Tr	BA2	GY
17	SCHLESWIG-HOLSTEIN	15187t	97	19.0k	142.0m	900P	294C	30Lr	BA2	GY
18F	SEA CORONA	12110t	72	17.0k	137.5m	12P		106T	A	NO
19p	SØNDERHO	93t	62	10.0k	26.3m	199P	0C	0L	-	DK
20	SPODSBJERG	958t	72	13.0k	67.3m	225P	40C	9L	BA	DK
21	SVEALAND	25026t	99	21.5k	186.0m	327P	164C	170T	A	MA
22	THOR SYDFYEN	1479t	78	12.0k	71.0m	292P	48C	9L	BA	DK
23	TYCHO BRAHE	11148t	91	14.0k	111.2m	1250P	240C	35Lr	BA	DK
24	URD	13144t	81	17.0k	171.0m	100P	-	104T	AS	DK

ASK Built by Nouvi Cantieri Aquania SpA, Venice, Italy as the LUCKY RIDER, a ro-ro freight ferry, for *Delpa Maritime* of Greece. In 1985 she was acquired by *Stena Line* and renamed the STENA DRIVER. Later that year she was acquired by *Sealink British Ferries* and renamed the SEAFREIGHT FREEWAY to operate freight-only services between Dover and Dunkerque. In 1988 she was sold to *SOMAT* of Bulgaria for use on *Medlink* services in the Mediterranean and renamed the SERDICA. In 1990 she was sold and renamed the NORTHERN HUNTER. In 1991 she was sold to *Blæsbjerg* of Denmark, renamed the ARKA MARINE and chartered to *DSB*. She was then converted into a ro-pax vessel, renamed the ASK and introduced onto the Århus – Kalundborg service. Purchased by *Scandlines AS* of Denmark in 1997. In 1999 she was, after some modification, transferred to *Scandlines Euroseabridge* and placed on the Travemünde – Klaìpeda route. In 2000 she was transferred to the Rostock – Liepaja route. Lengthened by 20m in 2001 and, in late 2001, chartered to *Nordö Link* to operate between Travemünde and Malmö. In late 2002 replaced by the FINNARROW and returned to *Scandlines*. She was transferred to the Rostock – Trelleborg route whilst the MECKLENBURG-VORPOMMERN was being rebuilt. She was then transferred to the Kiel – Klaìpeda route. In 2003 chartered to *Scandlines AB* to operate on the Trelleborg – Travemünde route. In April 2005 the charter ended and she returned to *Scandlines AG*. Initially due to replace the FELLOW on the Nynäshamn– Ventspils route during her annual refit. After that her use is undecided.

DEUTSCHLAND Train/vehicle ferry built by Van der Giessen-de Noord, Krimpen aan den IJssel, Rotterdam, Netherlands for *DFO* for the Puttgarden – Rødby service. During winter 2003/4 extra mezzanine deck added for cars.

FELLOW Built by Oy Wärtsilä Ab, Turku, Finland for *Finncarriers* as the FINNFELLOW. In 1989 transferred to *FinnLink*. In 2002 chartered to *VV-Line* and renamed the FELLOW. In 2003 *VV-Line* went into liquidation and she was chartered to *Scandlines*, who took over the Nynäshamn – Ventspils service.

FENJA Vehicle ferry built by Morsø Værft, Nykøbing Mors, Denmark for *SFDS A/S* for the Esbjerg – Nordby service.

FRIGG SYDFYEN Vehicle ferry built by Svendborg Skibsværft A/S, Svendborg, Denmark for *Sydfyenske Dampskibsselskab (SFDS)* of Denmark for the service between Spodsbjerg and Tårs. In 1996, this company was taken over by *DSB Rederi* and the company is now *Scandlines Sydfynske A/S*.

HAMLET Road vehicle ferry built by Finnyards, Rauma, Finland for *Scandlines* (50% owned by *Scandlines AG* and 50% owned by *Scandlines AB* of Sweden) for the Helsingør – Helsingborg service. Sister vessel of the TYCHO BRAHE but without rail tracks.

HOLGER DANSKE Built by Aalborg Værft A/S, Aalborg, Denmark as a train/vehicle ferry for *DSB* for the Helsingør – Helsingborg service. In 1991 transferred to the Kalundborg – Samsø route (no rail facilities). In 1997, transferred to subsidiary *SFDS A/S*. Withdrawn at the end of November 1998 when the service passed to *Samsø Linien*. In 1999 began operating between Rødby and Puttgarden as a road freight only vessel, carrying, among others, loads which cannot be conveyed on passenger vessels.

KRONPRINS FREDERIK Train/vehicle ferry built by Nakskov Skibsværft A/S, Nakskov, Denmark for *DSB* for the Nyborg – Korsør service. Withdrawn in 1997. After conversion to car/lorry ferry, she was transferred to the Gedser – Rostock route (no rail facilities).

MECKLENBURG-VORPOMMERN Train/vehicle ferry built by Schichau Seebeckswerft, Bremerhaven, Germany for *DFO* for the Rostock – Trelleborg service. During winter 2002/3 modified to increase freight capacity and reduce passenger capacity.

MENJA Built by Morsø Værft, Nykøbing Mors, Denmark for *SFDS A/S* for the Esbjerg – Nordby service.

ODIN SYDFYEN Vehicle ferry built by Svendborg Værft, Svendborg, Denmark for *Sydfyenske Dampskibsselskab (SFDS)* (now *Scandlines Sydfynske A/S*) of Denmark for the service between Spodsbjerg and Tårs.

PETERSBURG Built by Mathias Thesen Werft, Wismar, Germany (DDR) as the MUKRAN for *DSR* of Germany (DDR). In 1995 she was rebuilt to introduce road vehicle and additional passenger capacity and was renamed the PETERSBURG. She inaugurated the Travemünde service in 1995 but is now used on the Sassnitz – Klaìpeda service. This service was operated jointly with the *Lisco* vessel KLAIPEDA, a sister vessel which has not been converted to ro-pax format. In 2001 she was transferred to the Kiel – Klaìpeda service, replacing the sister vessel GREIFSWALD whose charter was ended. In 2003 replaced by the SVEALAND and transferred to the *Amber Line* Karlshamn – Liepaja route.

PRINS JOACHIM Train/vehicle ferry, built by Nakskov Skibsværft A/S, Nakskov, Denmark for *DSB* for the Nyborg – Korsør service. Withdrawn in 1997 and laid up. During winter 2000/2001 modified in same way as KRONPRINS FREDERIK and transferred to the Gedser – Rostock route.

PRINS RICHARD, PRINSESSE BENEDIKTE Train/vehicle ferries, built by Ørskov Christensen Staalskibsværft A/S, Frederikshavn, Denmark for Scandlines A/S for the Rødby – Puttgarden service. During winter 2003/4 extra mezzanine deck added for cars.

SASSNITZ Train/vehicle ferry built by Danyard A/S, Frederikshavn, Denmark for *Deutsche Reichsbahn*. In 1993 ownership transferred to *DFO*. Used on the Sassnitz – Trelleborg service.

SCHLESWIG-HOLSTEIN Train/vehicle ferry built by Van der Giessen-de Noord, Krimpen aan den IJssel, Rotterdam, Netherlands for *DFO* for the Puttgarden – Rødby service. During winter 2003/4 extra mezzanine deck added for cars.

SEA CORONA Built by Rauma Repola Oy, Rauma, Finland as the ANTARES for *Finska Ångfartygs A/B* of Finland and used on services between Finland and Germany. In 1975 she was chartered to *DG Hansa* of Germany and renamed the RHEINFELS. In 1977 she was sold to *Nedlloyd* of the Netherlands and renamed the NEDLLOYD ROCKANJE. In 1977 she was chartered to *Constellation Line* of the USA for services between the USA and Europe. In 1983 she was sold to *Kotka Line* of Finland, renamed the KOTKA LILY and used on their services between Finland, UK and West Africa. In 1985 she was chartered to *Jahre Line* of Norway, renamed the JALINA and operated freight services between Oslo and Kiel. Two years later, she returned to Baltic waters, being chartered to *Finncarriers* and renamed the FINNROVER. In 1988 she was chartered to *Kent Line*, renamed the SEAHORSE and used on their Dartford – Zeebrugge service. In 1991 she was chartered to *DFDS* and in 1992 she was renamed the DANA CORONA. She was initially used on the service between the Immingham and Cuxhaven (Germany) but in 1995 she was transferred to the Fredericia –

Copenhagen – Klaìpeda (Lithuania) service. In 2001 the charter was ended and she returned to owners, being renamed the SEA CORONA. She was chartered to *Scandlines AG* and placed on the *Scandlines Balticum Seaways* Århus – Aabenraa – Klaìpeda route.

SØNDERHO Passenger only ferry built by Esbjerg Jernstøberi og Maskinfabrik, Esbjerg, Denmark for *Post & Telegrafvæsenet* (Danish Post Office). In 1977 taken over by *DSB*. Used on extra peak sailings and late night and early morning sailings between Esbjerg and Nordby.

SPODSBJERG Vehicle ferry built by Husumer Schiffswerft AG, Husum, Denmark as the ÆRØ-PILEN for *Øernes D/S* of Denmark for services to the island of Ærø. In 1974 sold to *Sydfyenske Dampskibsselskab (SFDS)* for the service between Spodsbjerg and Tårs.

SVEALAND Ro-pax vessel built by C N Visentini di Visentini Francesco & C Donada, Italy as the ALYSSA for *Levantina Transporti* of Italy for charter. She was initially chartered to *CoTuNav* of Tunisia for service between Marseilles, Genoa and Tunis and in 2000 to *Trasmediterranea* of Spain for service between Barcelona and Palma de Majorca. In 2001 chartered to *Stena Line Scandinavia AB*, renamed the SVEALAND and placed as second vessel on the *Scandlines AB* freight only Trelleborg – Travemünde service. In 2003 sub-chartered to *Scandlines AG* and placed on the Kiel – Klaìpeda route, replacing the ASK and PETERSBURG.

THOR SYDFYEN Vehicle ferry built by Dannebrog Værft, Århus, Denmark for *Sydfyenske Dampskibsselskab (SFDS)* of Denmark (now *Scandlines Sydfynske A/S*) for the service between Spodsbjerg and Tårs. In 1998 she was transferred to the Fynshav – Bøjden route.

TYCHO BRAHE Train/vehicle ferry, built by Tangen Verft A/S, Tomrefjord, Norway for *DSB* for the Helsingør – Helsingborg service.

URD Built by Nouvi Cantieri Aquania SpA, Venice, Italy as the EASY RIDER, a ro-ro freight ferry, for *Delpa Maritime* of Greece and used on Mediterranean services. In 1985 she was acquired by *Sealink British Ferries* and renamed the SEAFREIGHT HIGHWAY to operate freight-only service between Dover and Dunkerque. In 1988 she was sold to *SOMAT* of Bulgaria for use on *Medlink* services in the Mediterranean and renamed the BOYANA. In 1990 she was sold to *Blæsbjerg* of Denmark, renamed the AKTIV MARINE and chartered to *DSB*. In 1991 she was converted into ro-pax vessel, renamed the URD and introduced onto the Århus – Kalundborg service. Purchased by *Scandlines* in 1997. Withdrawn at the end of May 1999 and, after modification, transferred to the *Balticum Seaways* (later *Scandlines Balticum Seaways*) Århus – Aabenraa – Klaìpeda route. In 2001 lengthened moved to the Rostock – Liepaja route.

SCANDLINES (SWEDEN)

THE COMPANY *Scandlines AB* (formerly *SweFerry*) is a Swedish company, a subsidiary of *Stena Line Scandinavia AB* of Sweden.

MANAGEMENT Managing Director Gunnar Blomdahl.

ADDRESS Knutpunkten 43, 252 78 Helsingborg, Sweden.

TELEPHONE Administration +46 (0)42-18 60 00, **Reservations *Helsingborg*** +46 (0)42-18 61 00, ***Trelleborg*** +46 (0)410-65 000.

FAX Administration +46 (0)42-18 60 49, **Reservations *Helsingborg*** +46 (0)42-18 74 10, ***Trelleborg*** +46 (0)410-65 001.

INTERNET Email kundservice@scandlines.se **Website** www.scandlines.se *(Swedish, English)*

ROUTES OPERATED Conventional Ferries Helsingborg (Sweden) – Helsingør (Denmark) (25 mins; *(1)*; every 20 mins), Trelleborg (Sweden) – Rostock (Germany) (6 hrs; *(3)*; 4 per day), Trelleborg – Sassnitz (Germany) (3 hrs 30 mins; *(4)*; 3 per day), Trelleborg (Sweden) – Travemünde (Germany) (8 hrs; *(2)*; 1 per day – freight-only). All routes are joint with *Scandlines AG* except Trelleborg – Travemünde.

1	AURORA AF HELSINGBORG	10918t	92	14.9k	111.2m	1250P	240C	34Lr	BA	SW
2	GÖTALAND	18060t	73	18.5k	183.1m	400P	118C	94Tr	AS2	SW

3	SKÅNE	42558t	98	21.0k	200.2m	600P	-	240Tr	AS2	SW
4	TRELLEBORG	20028t	82	21.0k	170.2m	900P	108C	48Tr	A2	SW

AURORA AF HELSINGBORG Train/vehicle ferry built by Tangen Verft A/S, Tomrefjord, Norway for *SweFerry* for *ScandLines* joint *DSB/SweFerry* service between Helsingør and Helsingborg. Owned by *Aurora 93 Trust* of the USA and chartered to *Scandlines*.

GÖTALAND Train/vehicle ferry built by A/S Nakskov Skipsværft, Nakskov, Denmark for *Statens Järnvägar (Swedish State Railways)* for freight services between Trelleborg and Sassnitz. In 1990 transferred to *SweFerry*. In 1992 modified to increase passenger capacity in order to run in passenger service. She is was on the Trelleborg – Rostock service until autumn 1998 when she was replaced by the SKÅNE. She then inaugurated a new freight-only Trelleborg – Travemünde service.

SKÅNE Train/vehicle ferry built by Astilleros Españoles, Cadiz, Spain for an American trust and chartered to *Scandlines*. She is used on the Trelleborg – Rostock service.

TRELLEBORG Train/vehicle ferry built by Öresundsvarvet Ab, Landskrona, Sweden for *Svelast* of Sweden (an *SJ* subsidiary). In 1990 ownership transferred to *SweFerry*. She is used on the Trelleborg – Sassnitz service.

SCF ST PETERSBURG LINE

THE COMPANY *SCF St Petersburg Line* is a Russian company, a subsidiary of *Sovcomflot*. Although the ships used are basically freighters, they market themselves as a passenger as well as freight line.

MANAGEMENT Managing Director Alexander I Dashkov.

ADDRESS 122. 63, Zhukovskogo Str., St Petersburg, 193036, Russian Federation.

TELEPHONE +7 (812) 324-7260.

FAX +7 (812) 324-7263.

INTERNET Email info@scflines.com **Website** www.scflines.com *(English, Russian, Germany)*

ROUTE OPERATED St Petersburg (Russia) – Kiel (Germany) (61 hours; *(1,2)*; 2 per week).

1	BREANT	5197t	79	13.5k	110.5m	7P	-	52T	A	UK
2	HAMBURG	8454t	77	14.5k	123.6m	12P	555C	64T	A	MA

BREANT Built by Karlskrona Varvet Ab, Karlskrona, Sweden for Lennart Kihlberg of Sweden. Operated by *Uniship Distribution* between Iggesund (Sweden) and Amsterdam (Netherlands). In 1981 sold to *Ivar Lundh & Co AB* and in 1995 sold to *Brax Shipping* of Sweden. In 1988, fitted with internal ramp between main deck and upper deck. In 1996 she was chartered to *Transfennica* and operated on services from Finland to Germany and the UK. In 2000 sold to *Oldhurst Ltd* of the British Virgin Islands. In 2001 chartered to *Bow Marine Oy* and operated between Helsinki – Hamina and Tallinn and Bremen and Bremerhaven. In 2002 she was chartered to *SCF Viborg Line* for their service between Viborg (Russia) and Kiel. In 2003 the service was transferred to St Petersburg and the name of the company changed to *SCF St Petersburg Line*. In 2004 she was sold to *Rederi Ab Lillgaard* of Finland. In November 2004 she was chartered for one week only to *Channel Freight Ferries*. She was then chartered to *Dart Line* to operate on the Dartford – Dunkerque service until the service ended about three weeks later. In December 2004 chartered to *SCF St Petersburg Line* again.

HAMBURG Built by Hollming Oy, Rauma, Finland as the MEKHANIK FEDOROV for *Latvian Shipping* of the former Soviet Union. In 1991 renamed the MEHANIKIS FJODOROVS (the same name but in Latvian rather than Russian). She was generally used on services from Latvia to North West Europe. In 1999 she was chartered to *CCTL*, renamed the CCTL HAMBURG and placed on their Hamburg – Hull service. Service switched to Killingholme in November 2000. In 2002 the charter ended, she was renamed the HAMBURG and she was employed on a number of short term charters. In 2003 chartered to *Mann Lines* and used on their Paldiski – Kiel – Århus – Gdansk – Paldiski service. In 2004 chartered to *SCF St Petersburg Line*.

SEA ADMINISTRATION OF ST PETERSBURG

THE COMPANY *Sea Administration of St Petersburg* is an agency of the Russian government.

ADDRESS St Petersburg (agents) Port Authority, Port of Saint Petersburg, Gapsalskaya str, 10, Saint Petersburg, 198035, Russian Federation. **Kaliningrad (agents)** Baltfinn-Kaliningrad Maritime Agency, Suvorova str, 45, Kaliningrad, 236039, Russia.

TELEPHONE St Petersburg (agents) +7 (812) 380-7094, **Kaliningrad (agents)** +7 0112 728-401 – agency dept (English speaking), +7 0112 728-404 – booking/ticket office (Russian speaking only), +7 0112 756-803 – cargo/truck forwarder (Russian speaking only).

FAX St Petersburg (agents) +7 (812) 380-7095, **Kaliningrad (agents)** +7 0112 728402.

INTERNET Email Saint Petersburg povg@mail.pasp.ru maai@mail.pasp.ru **Kaliningrad** office@baltfinn **Website** www.baltfinn.ru *(English)*

ROUTE OPERATED St Petersburg (Russia) – Baltijsk (Kaliningrad, Russia) (36 hrs; *(1)*; 1 per week).

1	GEORG OTS	12549t	80	17.0k	136.8m	341P	107C	21T	BA	RU

GEORG OTS Built by Stocnia Szczecinska im A Warskiego, Gdansk, Poland for *Estonian Shipping Company (ESCO)*, then of the USSR, to operate between Tallinn and Helsinki. Later chartered to *Tallink* (which was at the time partly owned by *ESCO*). Rebuilt in 1993 to increase car capacity from 14 to 110 and bring up to modern standards. In 2000 charter ended and she was returned to *ESCO*. In 2002 sold to *Sea Administration St Petersburg* and placed on the St Petersburg – Baltijsk service.

SEAWIND LINE

THE COMPANY *SeaWind Line* is a Swedish private sector company owned by *Silja Service Oy*.

MANAGEMENT Managing Director Mats Rosin **Sales Manager** Anders Levin.

ADDRESS Linnankatu 84, 20100 Turku, Finland.

TELEPHONE Administration & Reservations +358 (0)2 2102 800.

FAX Administration & Reservations +358 (0)2 2102 810.

INTERNET Website www.seawind.fi *(English, Swedish, Finnish)*

ROUTE OPERATED Stockholm (Sweden) – Långnäs (Åland) – Turku (Finland) (10 hrs 45 mins; *(1,2,3)*; 2/3 per day), Helsinki – Tallinn) (3 hrs 15 mins; *(3)*; 2/3 per day) (not summer peak).

1	SEA WIND	15879t	71	18.0k	154.9m	260P	60C	88Tr	BAS	SW
2	SKY WIND	16925t	86	15.0k	188.4m	364P	-	130Tr	A	SW
3	STAR WIND	13788t	77	18.0k	158.6m	119P	100C	66Tr	A	SW

SEA WIND Train/vehicle ferry built by Helsingørs Skipsværft, Helsingør, Denmark as the SVEALAND for *Stockholms Rederi AB Svea* and used on the *Trave Line* Helsingborg (Sweden) – Copenhagen (Tuborg Havn) – Travemünde freight service. In 1981 she was sold to *TT-Saga Line* and operated between Travemünde and Malmö. In 1984 she was rebuilt to increase capacity and renamed the SAGA WIND. In 1989 she was acquired by *SeaWind Line*, renamed the SEA WIND and inaugurated a combined rail freight, trailer and lower priced passenger service between Stockholm and Turku.

SKY WIND Train/vehicle ferry built by Moss Rosenberg Værft, Moss, Norway as the ÖRESUND for *Statens Järnvägar* (*Swedish State Railways*) for the 'DanLink' service between Helsingborg and Copenhagen. Has 817 metres of rail track. Service ceased in July 2000 and vessel laid up. In 2001 sold to *Sea Containers Ferries* and, in 2002 was converted at Gdansk, Poland to passenger ferry. She was chartered to *SeaWind Line* and, in autumn 2002, replaced the STAR WIND on the Stockholm – Turku service.

STAR WIND Train/vehicle ferry built by Bergens Mekaniske Verksted A/S, Bergen, Norway as the ROSTOCK for *Deutsche Reichsbahn* of Germany (DDR). Used on freight services between Trelleborg and Sassnitz. In 1992 modified to increase passenger capacity in order to run in passenger service. In 1993 ownership transferred to *DFO* and in 1994 she opened a new service from Rostock to Trelleborg. In 1997 she was used when winds preclude the use of the new MECKLENBURG-VORPOMMERN. Following modifications to this vessel in late 1997, the ROSTOCK continued to operate to provide additional capacity until the delivery of the SKÅNE of *Scandlines AB*, after which she was laid up. In 1999 she was sold to *SeaWind Line*, renamed the STAR WIND and operated in freight-only mode. Initial plans to bring her passenger accommodation up to the standards required for Baltic night service were dropped. Later in 2002 replaced by the SKY WIND and transferred to the Helsinki – Tallinn route. She carried a limited number of ordinary passengers on some sailings. In May 2005 returned to the Stockholm – Turku service for summer peak.

THE COMPANY *Silja Oy Ab*, based in Finland, is a wholly owned subsidiary of *Sea Containers Ltd.*

MANAGEMENT CEO (and also acting "Line Director" Helsinki-Stockholm Route) Antti Pankakoski, **CFO** Steven Robson, **Senior Vice President, Passenger Services** Pekka J Helin, **Senior Vice President, Cargo Services** Johan Lindblad, **Senior Vice President, Turku-Stockholm Route** Mats Rosin, **Senior Vice President, Cruises and Baltic Routes & Business Development** Tuomas Routa, **Senior Vice President, Fleet Management** Christian Grönvall, **Senior Vice President, Corporate Affairs and External Communications** Tuomas Nylund.

ADDRESS Silja Oy Ab, FI-02060 SILJA, Finland.

TELEPHONE Administration *Finland* +358 (0)9 18041, **Reservations *Finland*** +358 (0)203 74 552, ***Sweden*** +46 (0)8-222 140.

FAX Administration & Reservations *Finland* +358 (0)9 180 4484, ***Sweden*** +46 (0)8-667 8681.

INTERNET Email info@silja.com **Website** www.silja.com *(English, Finnish and Swedish)*

ROUTES OPERATED Conventional Ferries *All year* Helsinki (Finland) – Mariehamn (Åland) – Stockholm (Sweden) (16 hrs; *(5,6)*; 1 per day), Turku (Finland) – Mariehamn (Åland) (day)/Långnäs (Åland) (night) – Stockholm (11 hrs; *(2,3)*; 1 or 2 per day). ***Summer Only*** St Petersburg (Russia) – Tallinn (Estonia) – Rostock (Germany) (38 hrs (St Petersburg – Tallinn, 12-15 hrs, Tallinn – Rostock, 24 hrs)); *(1)*; every 4 days). ***Winter, Spring and Autumn only*** Turku (Finland) – Mariehamn (day)/Långnäs (night) – Kapellskär (Sweden) (11 hrs; *(2)*; 1 per day). **Fast Ferry** Helsinki – Tallinn (1 hr 30 mins; *(7,8,9)*; up to 7 per day). **Cruise Service** Helsinki – Tallinn – Visby (Gotland), Helsinki – Riga (Latvia), Helsinki – St Petersburg and other destinations *(4)*.

1	FINNJET	32940t	77	31.0k	214.9m	1790P	374C	44T	BA	FI
2	SILJA EUROPA	59912t	93	21.5k	201.8m	3000P	400C	68T	BA	FI
3	SILJA FESTIVAL	34414t	85	22.0k	170.7m	2000P	400C	80T	BA	SW
4p	SILJA OPERA	25076t	92	21.0k	158.9m	1400P	0C	0T	-	SW
5	SILJA SERENADE	58376t	90	21.0k	203.0m	2641P	450C	70T	BA	FI
6	SILJA SYMPHONY	58377t	91	21.0k	203.0m	2641P	450C	70T	BA	SW
7»	SUPERSEACAT ONE	4462t	97	38.0k	100.0m	782P	175C	-	A	IT
8»	SUPERSEACAT THREE	4697t	99	38.0k	100.0m	800P	175C	-	A	IT
9»	SUPERSEACAT FOUR	4697t	99	38.0k	100.0m	752P	164C	-	A	IT

FINNJET Built by Oy Wärtsilä Ab, Helsinki, Finland for *Finnlines* to operate between Helsinki and Travemünde, replacing several more vessels with intermediate calls. Her exceptionally fast speed was achieved by the use of gas turbine engines. During winter 1981/82 she was equipped with diesel engines for use during periods when traffic did not justify so many crossings per week. Later the trading name was changed to *Finnjet Line*. In 1986 the company was acquired by *EFFOA* and the trading name changed to *Finnjet Silja Line*. In winter 1997/98 she operated between Helsinki and Tallinn (Muuga Harbour). In summer 1998 she operated a weekly Travemünde – Tallinn – Helsinki – Travemünde triangular service in addition to two weekly Travemünde – Helsinki round trips. In autumn 1998 she resumed operating between Helsinki and Tallinn and since summer 1999 she operated Helsinki – Tallinn – Rostock during the summer and Helsinki – Tallinn in the winter. From June until September 2004 operated Rostock – Tallinn – St Petersburg. Laid up during winter 2004/5. To resume the service in spring 2005.

SILJA EUROPA Built by Jos Meyer, Papenburg, Germany. Ordered by *Rederi AB Slite* of Sweden for *Viking*

work to be completed. In 2004 she was renamed the STENA FREIGHTER and entered service with *Stena Line* between Gothenburg and Travemünde.

STENA GERMANICA, STENA SCANDINAVICA Built by Stocnia i Komuni Paryski, Gdynia, Poland (STENA GERMANICA) and Stocznia im Lenina, Gdansk, Poland (STENA SCANDINAVICA) for *Stena Line* for the Gothenburg – Kiel service. Names were swapped during construction in order that the STENA GERMANICA should enter service first. There were originally intended to be four vessels. Only two were delivered to *Stena Line*. The third (due to be called the STENA POLONICA) was sold by the builders as an unfinished hull to *Fred. Olsen Lines* of Norway and then resold to *ANEK* of Greece who had her completed at Perama and delivered as EL VENIZELOS for service between Greece and Italy. The fourth hull (due to be called the STENA BALTICA) was sold to *A. Lelakis*, of Greece and was rebuilt as a cruise ship to be called REGENT SKY; however the project was never completed. She was broken up in 2004. During the summer period on some days, the vessel arriving in Gothenburg overnight from Kiel operates a round trip to Frederikshavn before departing for Kiel the following evening. During winter 1998/99 they were modified to increase freight capacity and reduce the number of cabins.

STENA JUTLANDICA Train/vehicle 'ro-pax' vessel built by Van der Giessen-de Noord, Krimpen aan den IJssel, Rotterdam, Netherlands for *Stena Line* to operate between Gothenburg and Frederikshavn. She was launched as the STENA JUTLANDICA III and renamed on entry into service.

STENA NAUTICA Built by Nakskov Skipsværft A/S, Nakskov, Denmark as the NIELS KLIM for *DSB (Danish State Railways)* for their service between Århus (Jutland) and Kalundborg (Sealand). In 1990 she was purchased by *Stena Rederi* of Sweden and renamed the STENA NAUTICA. In 1992 she was chartered to *B&I Line*, renamed the ISLE OF INNISFREE and introduced onto the Rosslare – Pembroke Dock service, replacing the MUNSTER (8093t, 1970). In 1993 she was transferred to the Dublin – Holyhead service. In early 1995 she was chartered to *Lion Ferry*. She was renamed the LION KING. In 1996 she was replaced by a new LION KING and renamed the STENA NAUTICA. During summer 1996 she was chartered to *Trasmediterranea* of Spain but returned to *Stena RoRo* in the autumn and remained laid up during 1997. In December 1997 she was chartered to *Stena Line* and placed on the Halmstad – Grenaa route. This route ended on 31st January 1999 and she was transferred to the Varberg – Grenaa route. During winter 2001/2 rebuilt to heighten upper vehicle deck and allow separate loading of load vehicle decks; passenger capacity reduced. On 16 February 2004 hit by the coaster JOANNA and holed. Due to return to service at the end of May 2004 after repairs at Gothenburg and Gdansk.

STENA NORDICA Built by Mitsubishi Heavy Industries, Shimonoeki, Japan as the EUROPEAN AMBASSADOR for *P&O Irish Sea* for the Liverpool – Dublin service. Service transferred to Mostyn in November 2001. Also operated between Dublin and Cherbourg once a week. In 2004 the Mostyn route closed and she was sold to *Stena RoRo*. Chartered to *Stena Line* to operate between Karlskrona and Gdynia and renamed the STENA NORDICA.

STENA SAGA Built by Oy Wärtsilä Ab, Turku, Finland as the SILVIA REGINA for *Stockholms Rederi AB Svea* of Sweden. She was registered with subsidiary company *Svea Line* of Turku, Finland and was used on *Silja Line* services between Stockholm and Helsinki. In 1981 she was sold to *Johnson Line* and in 1984 sold to a Finnish Bank and chartered back. In 1990 she was purchased by *Stena RoRo* of Sweden for delivery in 1991. In 1991 she was renamed the STENA BRITANNICA and took up service on the Hoek van Holland – Harwich service for Dutch subsidiary *Stena Line bv*, operating with a British crew. In 1994 she was transferred to *Stena Line's* Oslo – Frederikshavn route and renamed the STENA SAGA. During winter 2002/3 rebuilt to increase passenger capacity by 200.

STENA SCANRAIL Built by Van der Giessen-de Noord, Capelle an der IJssel, Rotterdam, Netherlands. Launched as the STENA SEATRADER for *Stena AB* and entered service as the SEATRADER. In 1976 she was lengthened and then demise chartered to *Bahjah Navigation* of Cyprus and renamed the BAHJAN. In 1981 charter ended and she was renamed the STENA SEARIDER. In 1983 chartered to *Snowdrop Shipping* of Cyprus and renamed the SEARIDER. The charter ended the following year and she renamed the name STENA SEARIDER. Later in 1984 she was renamed the TRUCKER and in 1985 again reverted to the name STENA SEARIDER. In 1987 she was converted to a train ferry to operate between Gothenburg and Frederikshavn, chartered to *Stena Line* and renamed the STENA SCANRAIL.

Menja *(Andrew Cooke)*

Stena Nordica *(Miles Cowsill)*

SUPERFAST FERRIES

THE COMPANY *Superfast Ferries* is a Greek car-passenger ferry operator, owned by the *Attica Group*.

MANAGEMENT Managing Director Alexander P Panagopulos, **Corporate Relations and Development Director** Yannis B Criticos, **General Manager, Germany of Attica Premium SA** Jens-Peter Berg.

ADDRESS *Greece* 157 C Karamanli Avenue, Voula 16673, Greece, ***Germany*** Superfast Ferries, Hermann-Lange-Strasse 1, Lübeck 23558, Germany.

TELEPHONE Administration *Greece* +30 (0)210 891 9500, **Administration *Germany*** +49 (0)451 880060, **Reservations** +49 (0)451 880060.

FAX Administration *Greece* +30 (0)210 891 9509, ***Germany*** +49 (0)451 8800630

INTERNET Email criticos@attica-group.com **Website** www.superfast.com (*English, German, Finnish, Russian*)

ROUTES OPERATED Rostock (Germany) – Hanko (Finland) (21 hrs; *(3,4)*; 1 per day), **Freight service** Rostock – Uusikaupunki, (Finland) (41 hrs; (*1,2*); 3 per week).

1F	MARIN	5972t	90	17k	114.4m	12P	-	92T	A	GR
2F	NORDIA	7935t	91	16.5k	123.0m	12P	-	86T	A	GR
3	SUPERFAST VII	30285t	01	27.9k	203.3m	717P	695C	110L	BA2	GR
4	SUPERFAST VIII	30285t	01	27.9K	203.3m	717P	695C	110L	BA2	GR

SUPERFAST VII, SUPERFAST VIII Built by Howaltswerke Deutsche Werft AG, Kiel, Germany for *Attica Enterprises* (now *Attica Group*) for use by *Superfast Ferries* between Rostock and Hanko.

MARIN Built by Hjørungavaag Shipyard, Hjørungavaag, Norway for *Hellesøy Skipsbyggeri AS* of Norway. In 1991 sold to *Rederi AB Lillgård*, of Finland and chartered to *Transfennica* of Finland for use on service between Finland and Northern Europe. In 1995 chartered to *Finncarriers Oy* of Finland and operated on Baltic services. In 1997 sold to *Oy Rettig AB Bore* of Finland, renamed the FINNBEAVER and chartered back. In 2002 renamed the BORE-MARI and chartered to *BornholmsTrafikken* to operate between Køge and Rønne (Bornholm). In 2004 chartered to *Stena Line* to operate between Gothenburg and Travemünde. Later chartered to *P&O Ferries* to operate between Hull and Zeebrugge and Rotterdam. In late 2004 sold to *Attica Group* and renamed the MARIN. In January 2005 inaugurated a new service between Rostock and Uusikaupunki.

NORDIA Built by Solheimsviken A/S at Bergen, Norway for *Oy Rettig AB* of Finland as the BORE NORDIA and used on *Bore Line* service between Turku – Kiel – Cuxhaven – Harwich. In 1992 chartered to Finncarriers and used on Baltic services. In 1997 she was renamed the FINNSEAL. In 2003 she was renamed the BORE NORDIA and chartered to *Transfennica*. In 2004 sold to *Attica Group* and renamed the NORDIA. In January 2005 inaugurated a new service between Rostock and Uusikaupunki.

TALLINK

THE COMPANY *AS Tallink Grupp*, (formerly *AS Hansatee Grupp* trading as *Tallink*), is an Estonian company owned by the *AS Infortar (*86.0%) and others. Services are marketed outside Estonia by *Tallink Finland Oy*.

MANAGEMENT Director, Tallink Finland Oy Keijo Mehtonen.

ADDRESS PO Box 195, 00181 Helsinki, Finland.

TELEPHONE Administration *Estonia* +372 (0)6 409800, ***Finland*** +358 (0)9 228211, **Reservations** +358 (0)9 228311.

FAX Administration & Reservations *Estonia* +372 (0)6 409810, ***Finland*** +358 (0)9 228 21242.

INTERNET Email keijo.mehtonen@tallink.fi **Websites** www.tallink.ee *(Estonian)* www.tallink.se *(Swedish)* www.tallink.fi *(Finnish, English)*.

ROUTES OPERATED Conventional Ferries Helsinki – Tallinn (3 hrs 30 mins; *(3,6)*; up to 3 per day), Stockholm – Långnäs (Åland) – Tallinn (14 hrs; *(5,12)*; daily), Kapellskär – Paldiski (9 hrs – 11 hrs; *(11)*; 1 per day). **Freight Ferries** Helsinki – Tallinn (3 hrs 30 mins; *(2)*; 1 per day), Kapellskär – Paldiski (9 hrs – 11 hrs; *(4)*; 1 per day). **Fast Ferries** Helsinki – Tallinn (1 hr 30 mins; *(7,8,9,10)*; up to 10 per day).

1	FANTAASIA	16630t	79	21.3k	136.1m	1700P	549C	46T	BA	ES
2F	KAPELLA	‡2794t	74	14.5k	110.1m	50P	-	42T	A	ES
3	MELOODIA	17955t	79	21.0k	138.8m	1500P	480C	52T	BA	ES
4F	REGAL STAR	15281t	00	17.5k	156.6m	100P	-	-	A	ES
5	REGINA BALTICA	18345t	80	21.3k	145.2m	1450P	500C	68T	BA	ES
6	ROMANTIKA	40000t	02	22.0k	193.8m	2178P	300C	82T	BA	ES
7»	TALLINK AUTOEXPRESS	4859t	95	32.0k	78.6m	586P	150C	-	BA	ES
8»	TALLINK AUTOEXPRESS 2	5419t	97	37.0k	82.3m	700P	175C	-	A	ES
9»	TALLINK AUTOEXPRESS 3	4000t	97	36.0k	95.0m	600P	155C	-	A	ES
10»	TALLINK AUTOEXPRESS 4	3750t	96	36.0k	95.0m	580P	155C	-	A	ES
11	VANA TALLINN	10002t	74	18.0k	153.7m	1500P	300C	44L	BAS	ES
12	VICTORIA I	40000t	04	22.0k	193.8m	2500P	300C	823T	BA	ES

FANTAASIA Built by Oy Wärtsilä Ab, Turku, Finland as the TURELLA for *SF Line* of Finland for the *Viking Line* Stockholm – Mariehamn – Turku service and later moved to the Kapellskär – Mariehamn – Naantali service. In 1988 she was sold to *Stena Line*, renamed the STENA NORDICA and placed onto the Frederikshavn – Moss (night) and Frederikshavn – Gothenburg (day) service. In 1996 the Frederikshavn – Moss service ceased and she was transferred to subsidiary *Lion Ferry* and renamed the LION KING. She operated between Halmstad and Grenaa. In December 1997 she was sold to *Tallink Line Ltd* of Cyprus and renamed the FANTAASIA. In February 1998, after substantial modification, she was placed on the *Tallink* service between Helsinki and Tallinn. In June 2002 moved to the Tallinn – Stockholm route, enabling a daily service to be reinstated. In 2004 inaugurated a new Helsinki – Tallinn – St Petersburg – Helsinki service. In 2005 chartered to *Algérie Ferries* of Algeria, to operate between Algeria and France.

KAPELLA Built by Kristiansands Mekaniske Verksted, Kristiansand, Norway for *A/S Larvik-Frederikshavnferjen* of Norway as DUKE OF YORKSHIRE. In 1978 she was chartered to (and later purchased by) *CN Marine* of Canada (from 1986 *Marine Atlantic*) and renamed the MARINE EVANGELINE. She was used on services between Canada, USA and Newfoundland. In 1992 she was chartered to *Opale Ferries* of France and inaugurated a new Boulogne – Folkestone freight service. In 1993 the company went into liquidation and the service and charter were taken over by *Meridian Ferries*, a British company. She was renamed the SPIRIT OF BOULOGNE. In spring 1995, *Meridian Ferries* went into liquidation and she returned to her owners, resuming the name MARINE EVANGELINE. After a period of lay up she was chartered to *Stena Sealink Line*. She spent the summer on the Newhaven – Dieppe service and was then transferred to the Stranraer – Larne (from November 1995 Stranraer – Belfast) route. She returned to Newhaven in summer 1996 but was laid up during most of 1997. In late 1997 she was chartered to *Hansatee* and inaugurated a new ro-pax service between Paldiski and Kapellskär. She was renamed the KAPELLA. In 2004 transferred to the Helsinki – Tallinn route.

MELOODIA Built by Jos Meyer, Papenburg, Germany as DIANA II for *Rederi AB Slite* for *Viking Line* services between Stockholm and Turku, Mariehamn, Kapellskär and Naantali. In 1992 sold to a *Nordbanken* and, in 1993, chartered to *TR-Line* of Germany (joint venture between *TT-Line* and *DSR*) for service between Trelleborg and Rostock. In 1994 sold to *ESCO* and chartered to *EstLine* and renamed the MARE BALTICUM. During winter 1994/95 she was completely renovated. In 1996, following the delivery of the REGINA BALTICA, she was chartered to *Tallink*, renamed the MELOODIA and placed on the Tallinn – Helsinki service.

REGAL STAR Initially built by Sudostroitelnyy Zavod Severnaya Verf, St Petersburg. Work started in 1993 (as a deep sea ro-ro) but was never completed. In 1999 the vessel was purchased, taken to Palumba SpA, Naples and completed as a short-sea ro-ro with accommodation for 80 drivers. In

Stena Danica *(Miles Cowsill)*

Huckleberry Finn *(John May)*

2000 she was delivered to *MCL* of Italy and placed on a route between Savona and Catonia. In September of that year she was chartered to *Grimaldi Ferries* and operated on a route Salerno – Palermo – Valencia. In late 2003 she was sold to *Hansatee Shipping* of Estonia and, in 2004, placed on the Kapellskär – Paldiski route, replacing the KAPELLA.

REGINA BALTICA Built by Oy Wärtsilä Ab, Turku, Finland as the VIKING SONG for *Rederi AB Sally* of Finland and used on the *Viking Line* service between Stockholm and Helsinki. In 1985 replaced by the MARIELLA of *SF Line* and sold to *Fred. Olsen Lines*. She was named BRAEMAR and used on services between Norway and Britain as well as Norway and Denmark. Services to Britain ceased in June 1990 and she continued to operate between Norway and Denmark. She was withdrawn in 1991 and sold to *Rigorous Shipping* of Cyprus (a subsidiary of *Fred. Olsen Lines*). She was chartered to the *Baltic Shipping Company* of Russia, renamed the ANNA KARENINA and inaugurated a service between Kiel and St Petersburg. In 1992 a Nynäshamn call was introduced. In 1996 the service ceased and she was returned to her owners and renamed the ANNA K. Later in 1996 she was sold to *Empremare Shipping Co Ltd* of Cyprus (a company jointly owned by *Nordström & Thulin* and *Estonian Shipping Company*), chartered to *EstLine* and renamed the REGINA BALTICA. In 2000 charter transferred to *Tallink*. Continues to operate between Stockholm and Tallinn. Purchased by *Tallink* in 2002.

ROMANTIKA Built by Aker Finnyards, Rauma, Finland for *Tallink Grupp* to operate for *Tallink* between Tallinn and Helsinki.

TALLINK AUTOEXPRESS Austal Ships Auto Express 79 catamaran ordered by *Sea Containers* and launched at Fremantle, Western Australia as the AUTO EXPRESS 96. On completion she was renamed the SUPERSEACAT FRANCE. However, due to a dispute between *Sea Containers* and the builders, delivery was not taken and it was announced that she was to be sold to *Stena Rederi* of Sweden, renamed the STENA LYNX IV and chartered to *Stena Line (UK), I*naugurating a Newhaven – Dieppe service in February 1996. This did not happen and she was instead sold to *DSB Rederi (*now *Scandlines Danmark A/S*) and, in summer 1996, she was chartered to *Cat-Link* and renamed the CAT-LINK III. In 1999 she was sold to *Tallink* and renamed the TALLINK AUTOEXPRESS. Operates between Tallinn and Helsinki.

TALLINK AUTOEXPRESS 2 Austal Ships Auto Express 82 catamaran built at Fremantle, Western Australia as the BOOMERANG for *Polferries* and used on the Swinoujscie – Malmö route. In autumn 1999 she withdrawn and it was anticipated that she would no longer operate for *Polferries*. However in summer 2000 she returned to the Swinoujscie – Malmö route. She was laid up again in the autumn and in May 2001 was sold to Tallink and renamed the TALLINK AUTOEXPRESS 2. Operates between Tallinn and Helsinki.

TALLINK AUTOEXPRESS 3 Fincantieri MDV1200 monohull vessel. Built as the PEGASUS TWO at La Spézia, Italy for *Ocean Bridge Investments* and, in summer 1997 and summer 1998, chartered to *Color Line* to operate between Larvik and Skagen. In 1999 chartered to *Plonin Linija* of Croatia to operate a Venice – Ravenna – Plomin service. In 2000 sold to *African & Affairs SAL* of Spain and placed on a route between Malaga (Spain) and Nador (Morocco) until arrested in Gibraltar later in the year. In December purchased by *DMF Investments* of Gibraltar and in 2001 renamed the SHANNON ALEXIS. Laid up during 2001 and 2002. In 2003 briefly chartered to the *US Military*. In 2004 sold to *Tallink* and renamed the TALLINK AUTOEXPRESS 3.

TALLINK AUTOEXPRESS 4 Fincantieri MDV1200 monohull vessel. Built by La Spézia, Italy for *Ocean Bridge Investments*. Entered service as the STENA PEGASUS for *Stena Line* between Newhaven and Dieppe. In autumn 1996 she was laid up and later named the PEGASUS ONE. In 1997 she was chartered to *Conferrys* of Venezuela and operated between Puerto la Cruz and Punta de Piedras. In 1999 she was laid up in Greece. In 2002 she was moved to Gibraltar and renamed the ST MATTHEW. In 2004 she was sold to *Tallink* and renamed the TALLINK AUTOEXPRESS 4.

VANA TALLINN Built by Aalborg Skipsværft A/S, Aalborg, Denmark as the DANA REGINA for *DFDS* and used on their Esbjerg – Harwich service until 1983 when she was moved to the Copenhagen – Oslo route. In 1990 she was sold to *Nordström & Thulin* of Sweden, renamed the NORD ESTONIA and used on the *EstLine* Stockholm – Tallinn service. In 1992 she was chartered to *Larvik Line* to operate as a second vessel between Larvik and Frederikshavn and renamed the THOR HEYERDAHL.

section **8** changes since ferries 2004

british isles and northern europe

DISPOSALS

The following vessels, listed in the *Ferries 2004 – British Isles and Northern Europe* have been disposed of – either to other companies listed in this book or others. Company names are as used in that publication.

AURORA *(Scandlines (Denmark & Germany))* In 2005 chartered to *DFDS Tor Line* to operate between Zeebrugge and Immingham.

ASK *(Scandlines (Sweden))* In April 2005 returned to *Scandlines (Denmark & Germany).*

BALTIC EIDER *(Finnlines)* In 2004 sold to *Morbihannaise et Nantaise* of France and renamed the MN EIDER.

BOA VISTA *(KystLink)* In 2004 charter terminated. Sold to *Barber Shipmanagement* and chartered to *HUAL* for deep sea trading.

BRAYE SPIRIT *(It Ferries)* Vessel never delivered and service did not start.

CFF SOLENT *(Channel Freight Ferries)* In 2004 charter ended. Reverted to her previous name of TANGO. In 2004 chartered to *DFDS Tor Line* to operate on the Gothenburg – Gdansk service; later operated on the Zeebrugge – Immingham service. In 2005 renamed the FAST INDEPENDENCE.

CHODZIEZ *(Seawheel)* In 2004 service ended. In 2005 briefly chartered to *DFDS Tor Line* and operated between Zeebrugge and Immingham.

DART 2 *(Dart Line)* In 2005 sold to *Grimaldi Lines* and renamed the DAKAR BRIDGE. She is operate a feeder service from Dakar (Senegal) to Monrovia (Liberia), connecting with the company's services from Northern Europe to West Africa.

DRONNING MARGRETHE II *(Scandlines (Denmark and Germany))* In 2005 sold to *Marwan Shipping & Trading Co* of the United Arab Emirates to operate between India and Iraq. Later sold for scrap.

DUC DE NORMANDIE *(Brittany Ferries)* In 2004 withdrawn and laid up at Gdansk. In 2005 sold to *TransEuropa Shipping* of Slovenia and renamed the WISTERIA. To receive a refit in Ostend and is and now has been chartered to *Ferrimaroc.*

EUROPEAN DIPLOMAT *(P&O Irish Sea)* In 2004 sold to *Celtic Link Ferries* and renamed DIPLOMAT.

EXPRESS *(P&O Ferries)* In 2005 transferred to *P&O Irish Sea* to operate between Larne and Cairnryan/Troon.

FINNARROW *(Nordö Link)* In 2005 transferred to *FinnLink.*

FINNCLIPPER *(FinnLink)* In 2005 transferred to *Nordö Link.*

GRIMA *(Shetland Islands Council)* In 2004 sold to *Manson Marine.*

GUTE *(Rederi AB Gotland)* In 2004 chartered to *Riga Sea Lines.*

HAMBURG *(Mann Lines)* In 2004 charter ended. Chartered to *SCF St Petersburg Line.*

Pride of Provence *(Miles Cowsill)*

Duc de Normandie *(Miles Cowsill)*

JENS KOFOED (*BornholmsTrafikken)* In 2004 sold to *Eckerö Linjen.* Due to be delivered in May 2005. Renamed ECKERÖ.

HOVERSPEED GREAT BRITAIN *(Emeraude Jersey Ferries)* In 2005 renamed the EMERAUDE GB. In 2005 transferred to *Sea Containers* subsidiary *Aegean SpeedLines* renamed the SPEEDRUNNER 1 and operated from Piraeus to Serifos, Sifnos, Milos and Paros.

KAHLBERG *(Polferries)* In 2005 sold to *RG-Line.*

KAPTAN BURHANETTIN ISIM *(Vent-Lines)* In 2004 sold to *Polish Steamship Company* and renamed the GRYF. Chartered to *Unity Line.*

KOADA *(Shetland Islands Council)* In 2004 sold to *Olaf Isbister* of Shetland for uses as a charter vessel.

LEHOLA *(Scandlines)* Charter ended at the end of 2004. chartered to *P&O Ferries* for service from Hull to Rotterdam and Zeebrugge.

LEMBITU *(Scandlines)* Charter ended at the end of 2004. Chartered to *Channel Freight Ferries.* Then chartered to *NorseMerchant Ferries,* then to *Seatruck Ferries* and renamed the CHALLENGE.

MASSILIA *(Dart Line)* In early 2005 sold to a limited partnership under control of the *Eidsiva Rederi* of Norway. In autumn 2005 to be chartered to *DFDS Tor Line* and renamed the TOR BELLONA.

MAX MOLS *(P&O Ferries)* In 2004 returned to *Mols-Linien.*

MIDNATSOL II *(Hurtigruten)* In 2005 sold to Croisières Transboréales of Canada to operate in the Gulf of St Lawrence.

NO 4 *(C Toms & Sons Ltd)* In 2004 broken up.

PASCARA JET *(Sea Containers Ferries*) In 2005 renamed the ZARA JET.

PIONEER *(Caledonian MacBrayne)* In 2004 sold to *Burness Corlett* to operate in Sao Tome and Principe (in the Gulf of Guinea) and renamed the BRENDA CORLETT.

Express *(Miles Cowsill)*

PRIDE OF CHERBOURG *(P&O Ferries)* in 2005 charter transferred to *Stena RoRo*. Renamed the STENA CHALLENGER and sub chartered to *Stena Line* to operate between Karlskrona and Gdynia.

PRIDE OF PROVENCE *(P&O Ferries)* In 2004 sold to *GA Ferries* of Greece and renamed the ALKMINI A. Used on services to Patmos, Leros, Kalymnos, Kos and Rhodes.

RÜGEN *(Scandlines (Denmark and Germany))* In 2005 sold to *Marwan Shipping & Trading Co* of the United Arab Emirates to operate between India and Iraq. Later sold for scrap.

SEACAT ISLE OF MAN *(Isle of Man Steam Packet Company)* In 2005 chartered to *Irish Sea Express* and reamed the SEA EXPRESS 1.

SEAWHEEL HUMBER *(Seawheel)* In 2004 service ended. In 2005 chartered to *Finnlines* to operate for *TransRussia Express*. Renamed the BALTIC EAGER.

SEAWHEEL RHINE *(Seawheel)* In 2004 sold to *Argo Systems FZE*. Renamed the SEAWHEE. Later sold to *Giannatos* (*Med Link Lines*) of Greece and renamed the MARIA G.

SKAGEN *(Color Line)* In 2005 sold to *El Salam Maritime* of Egypt.

SOLENT ENTERPRISE (*Gosport Ferry*) In 2005 sold to *Capital Pleasure Boats* of London and renamed SUNDANCE.

STENA ENVOY *(Stena RoRo)* In 2004 sold to *KystLink* and renamed the ENVOY.

STENA SHIPPER *(DFDS Tor Line)* In 2005 charter ended. Chartered to *K Line European Sea Highway Service* (car transporters) to operate between Germany, Finland and Russia. Renamed the VOLGA HIGHWAY.

SUPERSTAR EXPRESS *(P&O Irish Sea)* In 2004 charter ended.

TOR GOTHIA *(DFDS Tor Line)* In 2004 sold to the *Ulusoy Group* of Turkey and renamed the ULUSOY-6. Operated between Cesme (Turkey) and Trieste (Italy).

VILJA *(BornholmsTrafikken)* In 2004 charter ended.

NAME CHANGES

The following ships have been renamed without change of operator:

DIAMANT *(Sea Containers Ferries)* In 2004 renamed the SEACAT DIAMANT.

LAGAN VIKING *(NorseMerchant Ferries)* In 2005 renamed the LIVERPOOL VIKING.

SEACAT FRANCE *(Sea Containers Ferries)* In February 2005 chartered to *Emeraude Jersey Ferries* and renamed the EMERAUDE FRANCE. In April returned to *Sea Containers Ferries*. May be renamed.

SOUND OF SLEAT *(Western Ferries)* In 2004 renamed the SLEAT.

COMPANY CHANGES

AWSR Shipping In 2004 name changed to *Foreland Shipping*.

Hugo Express In 2005 franchise to operate the service awarded to *Société Morbihannaise de Navigation* trading as *Manche Iles Express*.

It Ferries Service never started.

Saga Line In 2004 service ended after company got into financial difficulties.

Seawheel In 2004 service ended. Replaced by lo-lo service between Rotterdam and Goole.

Vent Lines In 2004 service ended following sale of the KAPTAN BURHANETTIN ISIM.

Mont St Michel *(Miles Cowsill)*